HOTEL

设计速递 酒店设计

精品文化工作室 / 编
王连双 张海云 康琪 / 译

大连理工大学出版社
Dalian University of Technology Press

图书在版编目(CIP)数据

酒店设计 : 汉英对照 / 精品文化工作室编 ; 王连双, 张海云, 康琪译. — 大连 : 大连理工大学出版社, 2012.5

(设计速递)

ISBN 978-7-5611-6831-8

Ⅰ. ①酒… Ⅱ. ①精… ②王… ③张… ④康… Ⅲ. ①饭店—建筑设计—作品集—世界 Ⅳ. ①TU247.4

中国版本图书馆CIP数据核字(2012)第059352号

出版发行：大连理工大学出版社
(地址：大连市软件园路80号　邮编：116023)
印　　刷：利丰雅高印刷(深圳)有限公司
幅面尺寸：225mm × 300mm
印　　张：15
插　　页：4
出版时间：2012年5月第 1 版
印刷时间：2012年5月第 1 次印刷
责任编辑：刘　蓉
责任校对：李　雪
封面设计：连　帅

ISBN 978-7-5611-6831-8
定　　价：210.00元

电 话：0411-84708842
传 真：0411-84701466
邮 购：0411-84703636
E-mail：designbooks_dutp@yahoo.cn
URL：http://www.dutp.cn

如有质量问题请联系出版中心：(0411) 84709246　84709043

Thoughts on the Design of Boutique Hotels

Hotels today are not only a lodging place where travelling guests have to stay. The emergence of boutique hotels represents the pursuit of fine life and personal interests by the elites to find a private space for relaxation and enjoyment. This concept originates from the Europe in 1970s and its specialty lies in the unusual choice of locations, the characteristic and fashionable decorations and furnishings and the detailed and human-oriented services, which provides the clients with private, fashionable and high-quality enjoyment. Boutique hotels of various styles emerge in cities of fashions all over the world, including styles of luxury, fashion, business, holiday, Southeast Asia, hydrotherapy, etc. Some of the boutique hotels with local characteristics have become local tourist sites. In Italy, one of the fashion capitals of the world, some boutique hotels are rebuilt out of ancient relics. With the nostalgic exterior appearance and the comfortable and modern interior design, sense of revival coexists with sense of the times, making the hotels a tourist spot that no travelers will miss. Except for the ornamental value, boutique hotels possess another functional value among others which helps to wipe out worries. Some of the hydrotherapy-themed boutique hotels provide the guests with all kinds of body and mind relaxing images. Surrounded by romantic, elegant and honored atmosphere, it is like staying far from the metropolitan hustle and bustle and entering into a fairyland as arcadia, forgetting about the worries and enjoying the wonderful experience brought by hydrotherapy. The delightful surprise that a boutique hotel can provide may beyond common expectations, therefore, a fine boutique hotel can invite people to linger on and leave a deep impression on them.

Through the designers' different designs, the boutique hotels in different cultures have their own souls. In the past, the boutique hotels I've done, such as Renaissance Beijing Chaoyang Hotel, possesses a traceless elegance and the prime quality of luxury which touches you deeply. My design focuses on the unique, profound culture heritage and solemn royal demeanor of the Beijing city. The architecture style is featured by a simple exterior appearance and for the interior, classical romance and French style luxury are adopted which makes it one of the scenic spots worth seeing in Beijing. However, Renaissance Shanghai Pudong Hotel takes the theme of "seeking for Shanghai". By combining the Chinese ornamental style, the design endows the hotel with rich cultural connotations and behind the simple and fashionable design, it's full of humanistic expressions from the architecture appearance to the indoor space. The setting art of the furnishings of the hotel plays a decisive role, which displays cultural connotations of the hotel with fashion; in the level of art, it promotes the commercial and ornamental values which provide the high-end boutique hotel with more connotations.

Entering 21 century, the future quality products will hopefully be an investment and business proposition. The boutique hotels have become a fashion and top fashion brands such as ARMANI, BVLGARI and MISSONI have already built boutique hotels and the display of fashion and luxury has become more and more skillful. Most of the boutique hotels are located in Italy, the capital of global luxury and New York and Dubai are also hot places of fashion. Hopefully, China will become the target location of the next batch of boutique hotels of fashion. We believe in the near future, with the efforts of the designers, China will also have top boutique hotels with characteristics.

IVAN C. DESIGN LIMITED
Ivan Cheng

精品酒店设计的感想

现在的酒店已不仅仅是客人旅行时不得不住的地方，精品酒店的出现代表了精英阶层对于精致生活、个人趣味的追求，让人拥有放松心情、享受生活的私密空间。这一概念源于20世纪70年代的欧洲，其特色在于特别的选址、个性时尚的装饰陈设、注重细节的人性化服务，能够提供给客户私密、时尚的高品质享受。各种风格的精品酒店出现于世界时尚之都，风格包括奢华、时尚、商务、度假、东南亚、水疗等类型。有些富有地方特色的精品酒店已成为当地的旅游景点。在时尚之都意大利，部分精品酒店由古迹改建而来，怀旧的外表以及舒适现代的内装，复古感与时代感并存，让酒店成为旅行家必去的景点。除了欣赏价值，精品酒店最值得一提的便是让人忘却烦恼的功能价值。有些以水疗为主题的精品酒店，会让宾客体验到各种可以放松身心的意境，浪漫、优雅、尊贵的感受环绕四周，让人仿佛远离城市的喧嚣，进入了一个世外桃源般的仙境，忘却生活的烦恼，体会水疗带来的美妙感觉。精品酒店可以做到的，也许是想象之外的惊喜，因此，一家好的精品酒店可以让人流连忘返、记忆深刻。

处于不同文化之中的精品酒店通过设计者不同的表达，有其自己的灵魂。过去，我完成的精品酒店，如北京国航万丽酒店，便有着不露痕迹的优雅、深彻入骨的顶级奢华品质。我在设计时抓住了北京这个城市所独有的深厚文化底蕴和大气的皇家风范，外表简约的建筑风格，而内在则采用了古典的浪漫和法式的奢华，成为当地值得欣赏的景点之一；而上海淳大万丽酒店则是以“寻找上海”为主题，结合中式的装饰主义风格，赋予了酒店丰富的文化内涵，在简约和时尚设计的背后，从建筑外观到室内环境都充满着浓厚的人文风情。酒店陈设艺术起着举足轻重的作用，让酒店在时尚之中尽显文化内涵；在艺术的层面上，提升商业及欣赏价值，使高端精品酒店更具内涵。

进入21世纪，未来的精品产品很有希望作为一个投资和商业命题。精品酒店已变成潮流时尚的酒店，ARMANI、BVLGARI 、MISSONI等顶级时尚品牌已纷纷打造精品酒店，人们对时尚与奢华的演绎也越来越纯熟。大部分的时尚精品酒店都位于世界奢侈品之都意大利，纽约和迪拜也是热点，中国有望成为下一批时尚精品酒店落户的目标。相信在设计师们的努力下，不久的将来，中国也会拥有富有特色的顶级精品酒店。

IVAN C. DESIGN LIMITED
郑仕樑

Contents 目录

地点	设计师	设计公司
/ 深圳	/ 刘波	/ 刘波设计顾问（香港）有限公司

Wongtee V Hotel

皇庭V酒店

Wongtee V Hotel is financed and built by the well-known real estate enterprise, the Wongtee Group in Shenzhen. It is a five-star hotel and is the first one to create the theme of uniqueness, fashion and art concepts. With the theme of fashion and avant-garde decorative style, the hotel combines both nature and art together in design, with aura throughout. The use of large quantities of post modern design elements and abstract elements create blurred fantasy and distinctive atmosphere for the hotel.

The hotel is located on Jintian Road in Futian downtown area. It dominates the rare estate in Shenzhen CBD business circle with superb location, convenient transportation and an access to the central supreme equipment. Located in the main building of Huanggang Business center, the hotel covers the area from 26th floor to 54th floor. It is a typical air hotel with almost 300 fashionable guest rooms. The rooms are comfortable and elegant with perfect privacy. More than half of its guest rooms offer the panoramic view of Shenzhen Bay, leaving people quite a different feeling.

The distinctive Chinese and western restaurants and bars offer a blend of Chinese and foreign food, leaving people good flavor and long-lasting aftertaste. Meanwhile, there are the highest air club in the city, multi-functional banquet hall and well-equipped fitness center. When entering the hotel, you may feel you were wandering in the air with blue sky and cloud around. You may feel dignified to overlook the beautiful scenery of this prosperous metropolis. It is in this central section of the city that people enjoy the unforgettable experience.

HOTEL

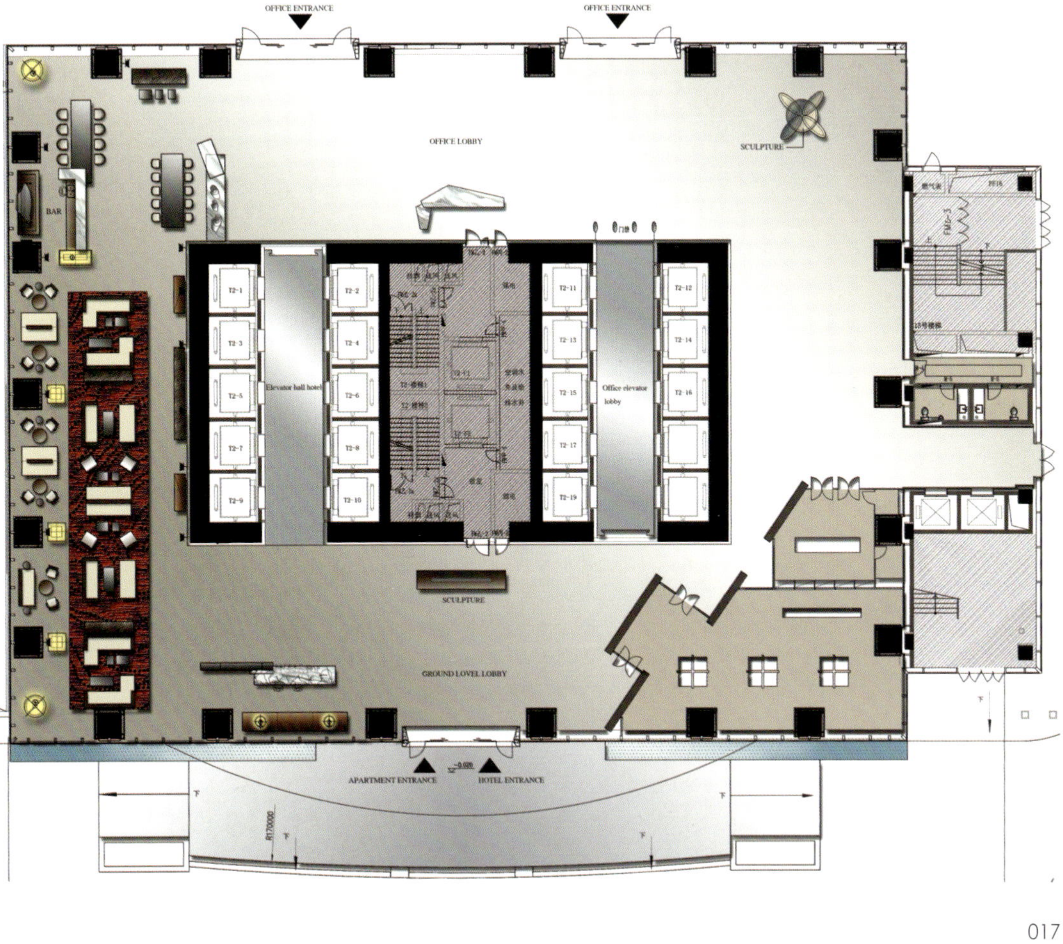

皇庭V酒店是由深圳知名地产企业——深圳皇庭集团按五星级标准投资打造的深圳市第一家以时尚独特、艺术概念为主题的酒店。酒店以时尚为主题，装修风格现代前卫。设计上融自然和艺术于一体，处处充满灵气，大量后现代设计艺术和抽象主义元素为酒店营造出迷离梦幻和个性鲜明的氛围。

酒店坐落于福田中心商业区金田路，执掌深圳CBD商业圈稀世地脉，地理位置优越，交通便利，尽享城市中央至尊配套。酒店位居皇岗商务中心主楼26层至54层，是典型的空中酒店，拥有近300间独具时尚概念的客房。房间舒适、典雅且私密性绝佳，半数以上的客房可将深圳湾滨海景观尽收眼底，给人与众不同的感受。

各具特色的中餐厅、西餐厅和酒吧，荟萃中外美食佳酿，让人齿颊留香、回味悠长。同时，酒店还拥有全城最高的空中俱乐部、功能齐全的宴会厅以及配套完备的康体中心，置身其中，犹如云中漫步，畅享蓝天白云，傲然俯瞰繁华都市美景，在城市之央享受难忘的旅程。

地点	面积	设计师	参与设计	设计公司	主要材料	摄影师
/ 深圳	/ 25000m²	/ 琚宾	/ 张轩崇、许金花、石燕、尹芮、谭琼妹	/ HSD水平线空间设计	/ 灰木纹、科技木、艺术地毯	/ 孙祥宇

Ascott Melody Hotel

雅士阁美伦酒店

Ascott Melody Hotel is located in Shekou, Shenzhen, a bustling and thriving place full of amiable life atmosphere, but it's not messy. For the reason that most of the guests are foreigners, it has a touch of exotic tastes. Shekou is near to the sea with rather humid climate, the wind carrying thick sea smells, and it is full of life atmosphere, but it feels that you are on a journey, which is destined to bring unique brilliance.

The exquisite but not abrupt architecture of the hotel echoes with the building community in the proximity. The interior design follows the feel of the hotel appearance. The modeling of the lobby ceiling and the facade structure out of the hotel remain unified, winding, soft and robust which seem in disorder but actually in harmony. Wood-like aluminum highlights the feeling of massiness and quality, and at the same time it maintains the feeling of relaxation and comfort reflecting the theme of business and holiday.

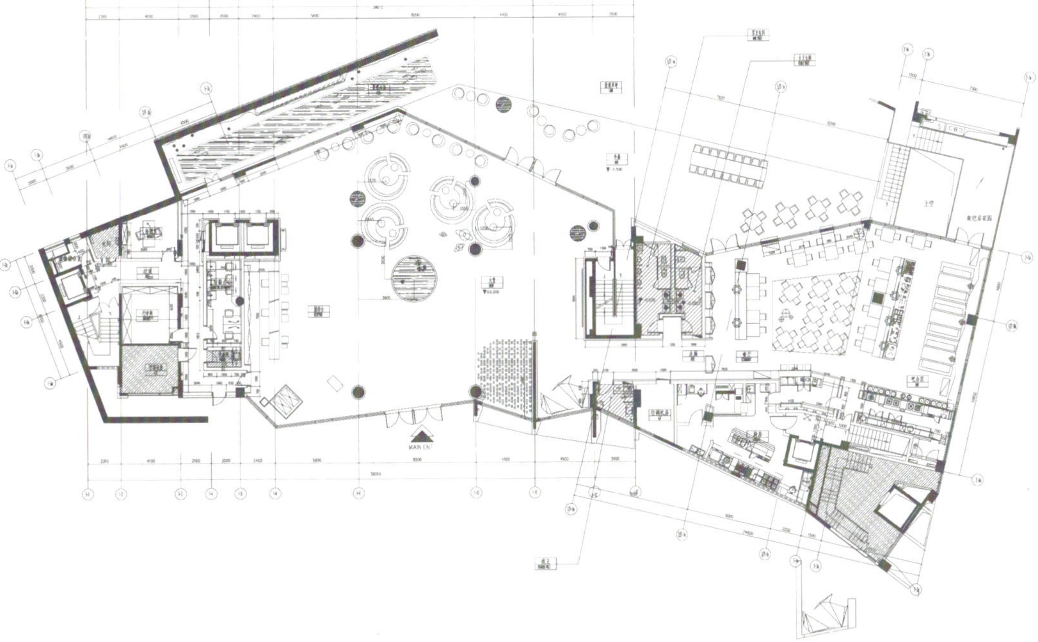

雅士阁美伦酒店位于深圳蛇口，热闹、繁华，但不杂乱，熙攘中带着点生活味的可爱气息。因为居住的外国人较多，因此沾着点异域的情调。蛇口临海，有点潮，风中海味很浓，生活气息很重，却感觉很“旅途”，这样的一种感觉注定带给人们别样的精彩。

酒店的建筑远远呼应着周边的建筑群落，既考究又不突兀。室内设计延续酒店外观的感受，大堂天花的造型与建筑外立面形体保持一致，蜿蜒折回，柔软而硬朗，似无序而协调。木状的铝质材料，在突显整个空间的厚重感和品质感的同时，保持了轻松感和舒适感，体现出商务与度假的主题。

地点	面 积	设计师	项目经理	设计公司	摄影师
/ 葡萄牙	/ 3200m²	/ Werner Franz, Alexander Plajer	/ Michael Bertram	/ Plajer & Franz Studio	/ Ken Schluchtmann

Porto Palácio Hotel

波尔图帕拉西奥酒店

The hotel is located in an 18-storey modern building in Porto, the ancient seaside city at the estuary of the Douro River. With the building first constructed in the 1970s, the present design is to complete the interior decoration of the hotel.

The design concept of this project is to create a subtle and elegant exterior with a high-end interior on the basis of the local design in Porto. In here, no matter where you go, you can get a glimpse at articles representing the history of Porto. The extensive use of large black images in monochrome background "bridge, terrain, ship, and gear" paired with ancient writings, these existences of the seemingly historical relics are used in modern patterns forming a photo album that leaves customers profound memories.

The designer uses round furniture as the most important interior embellishment, showing its artistic decoration style. As for the details, including the fine series of wallpaper, the curtains, the carpet, the fabric wallpaper, the well-framed images, and the modern chandeliers...all make the space an exhibition hall of art collection.

Porto Palácio
HOTEL

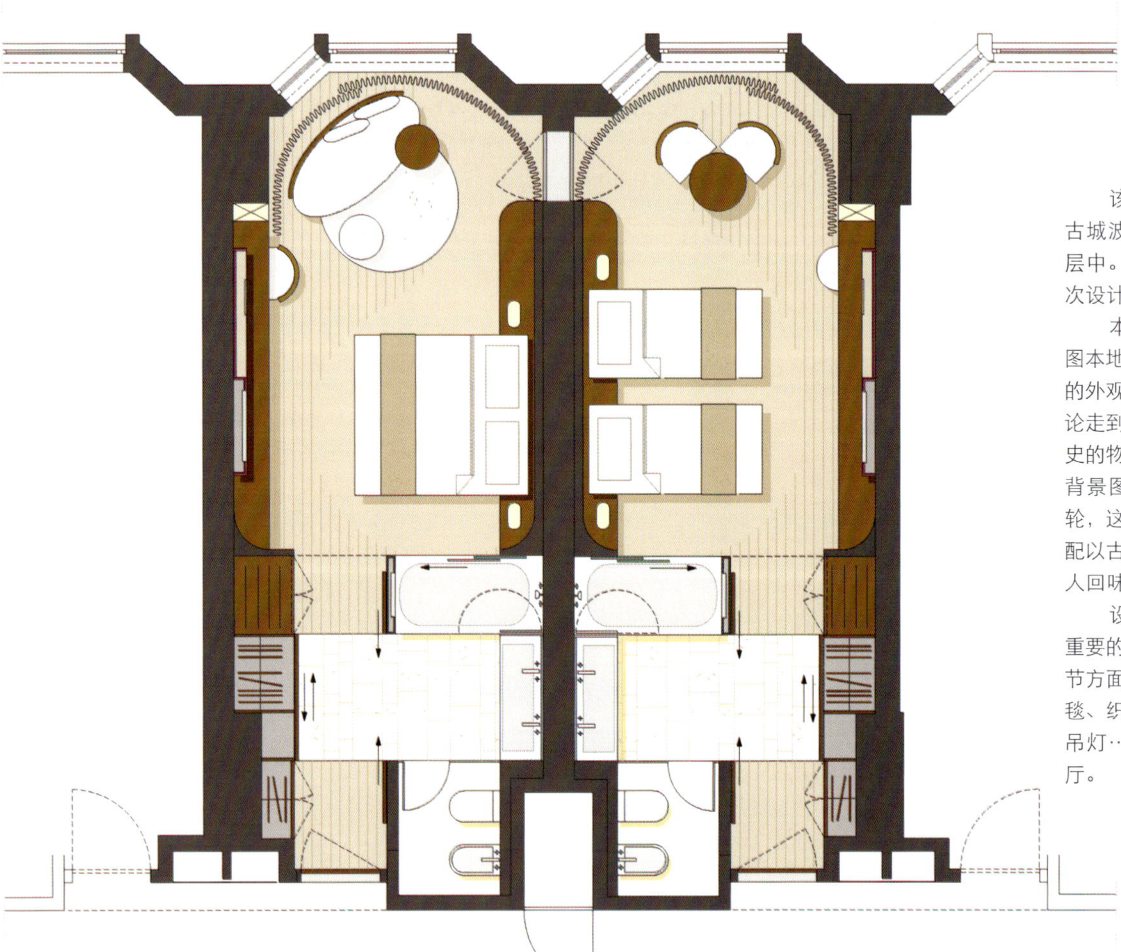

该酒店坐落于多罗河河口的海边古城波尔图，在一个18层高的现代高层中。大楼始建于20世纪70年代，本次设计要完成的是酒店的内部装修。

本案的设计理念是在已有的波尔图本地设计的基础上，创建微妙而优雅的外观和高端的室内设计。在这里，无论走到哪里，都能窥见代表着波尔图历史的物品。广泛使用的黑色的大型单色背景图像，例如桥梁、地形、船、齿轮，这些历史遗迹被用于现代图案中，配以古老的文字，就像是一本影集，让人回味。

设计师选用圆形家具作为室内最重要的点缀，来体现艺术装饰风格。细节方面，包括精品系列壁纸、窗帘、地毯、织物墙纸、精心装裱的画像、现代吊灯……让空间成为一个艺术收藏展厅。

地点	面 积	设计师	陈设设计师	材料设计师	设计公司	主要材料
/ 广州	/ 8500m²	/ 黄炽烽	/ 谭启开	/ 谭启开	/ J2-DESIGN顾问设计有限公司	/ 工艺不锈钢、波浪板、石材、镜面黑钢

Smart Hotel (Guangzhou)

时尚旅酒店(广州店)

Located by water, thrived because of water and excelled in possession of water, Guangzhou owns a long lasting watertown culture. This project just incorporates the water culture of Guangzhou and makes water as its theme, by adding appropriately metropolis modern elements, creating a characteristic atmosphere of modern watercity without losing any good taste.

Etching stainless steel is used for the background of the hotel lobby to draw the patterns of flowing water which appear to be more dynamic like the free running water with the cooperation of light and change of angles. Other elements such as the elevation materials on the wall, the handling of lamp strips and the binding off of the technics carefully modify the shapes of the running water. The coffee house takes the white-collar consumer group as a breakthrough point and dark colors are used more to show good taste in the low-key style. Wave panels are not only fashionable and elegant but also in concert with the theme of water. The checkered-mirror-finish black-steel ceiling shines upon the scene and adds agility to the space, glistening like the water surface.

广州滨江而立，因水而兴，拥水而优，水乡文化源远流长。本案正是融合了广州的水文化，以水为主题，并适当地添加了都市摩登元素，营造出既有特色又不失品位的现代水城氛围。

酒店大堂的背景运用不锈钢蚀花工艺，刻画出水流的图案，在灯光的配合下及角度的变换中更显动感，仿佛流水在恣意奔腾。墙上的立面材料、灯带的处理、工艺的收口都细致地修饰了水流的形态。咖啡厅则是以白领消费群为切入点，更多地运用沉稳的深色调，在低调中彰显品位。波浪板元素既时尚典雅又与水的主题遥相呼应，镜面黑钢方格天花映照着景物，增添了空间的灵动，犹如水面一般波光粼粼。

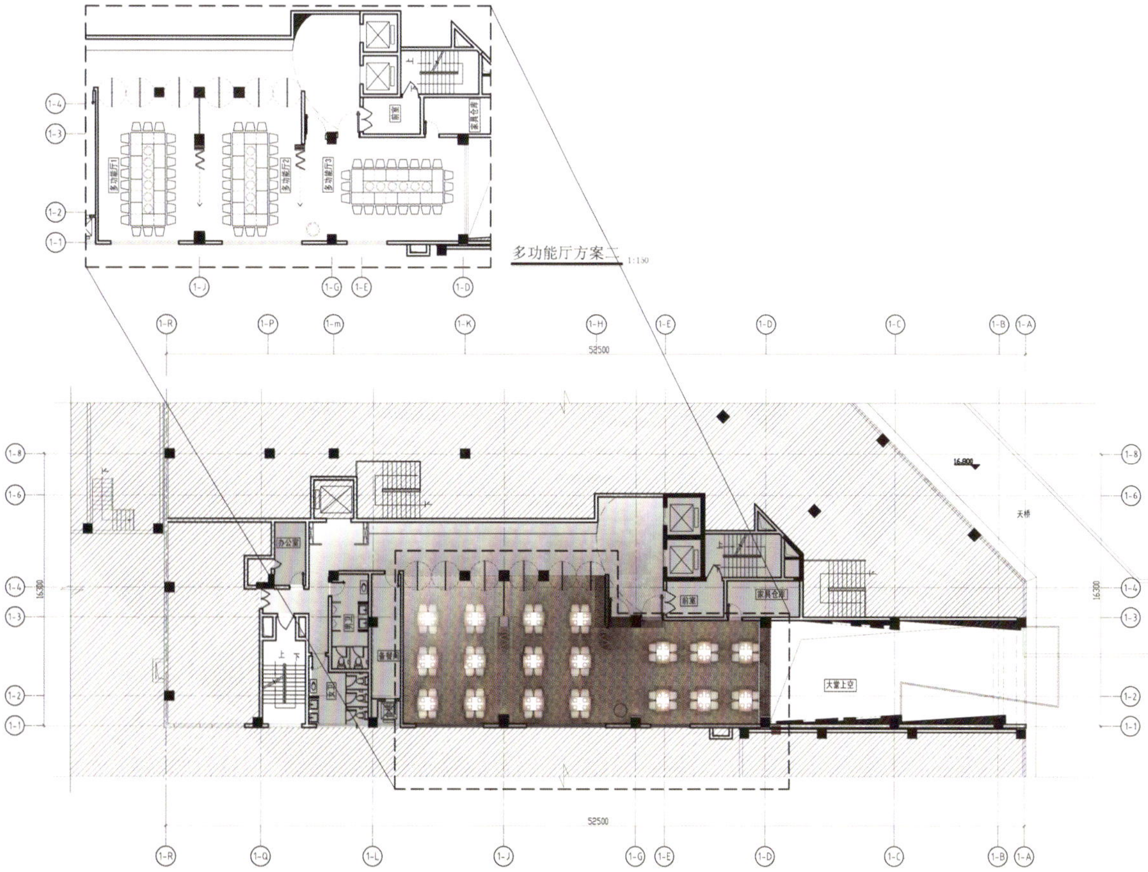

多功能厅方案二 1:150

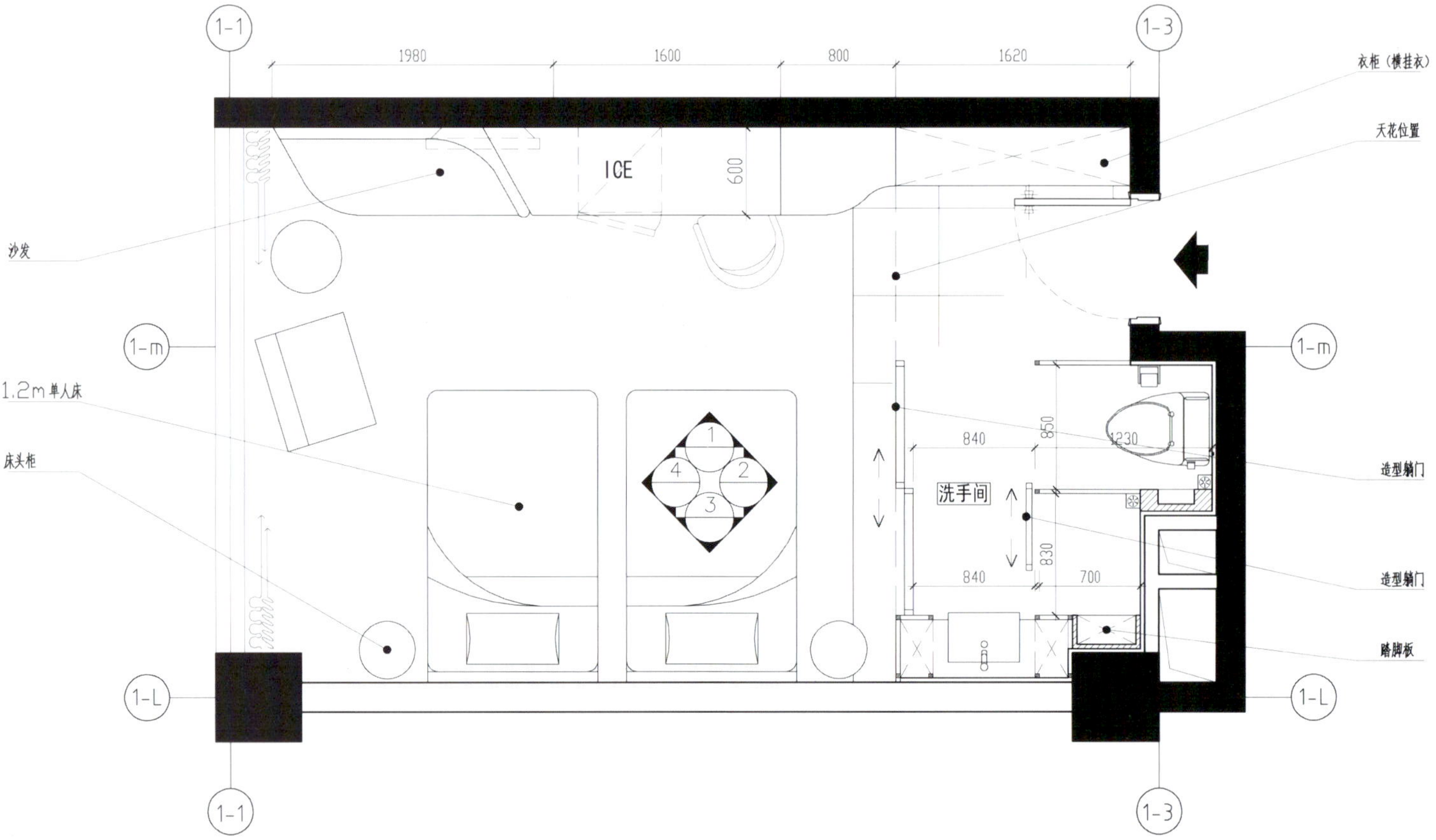
1-1
1-3
1980
1600
800
1620
衣柜（横挂衣）
天花位置
沙发
ICE
600
1-m
1.2m 单人床
床头柜
840
850
1230
洗手间
830
840
700
造型躺门
造型躺门
踏脚板
1-L

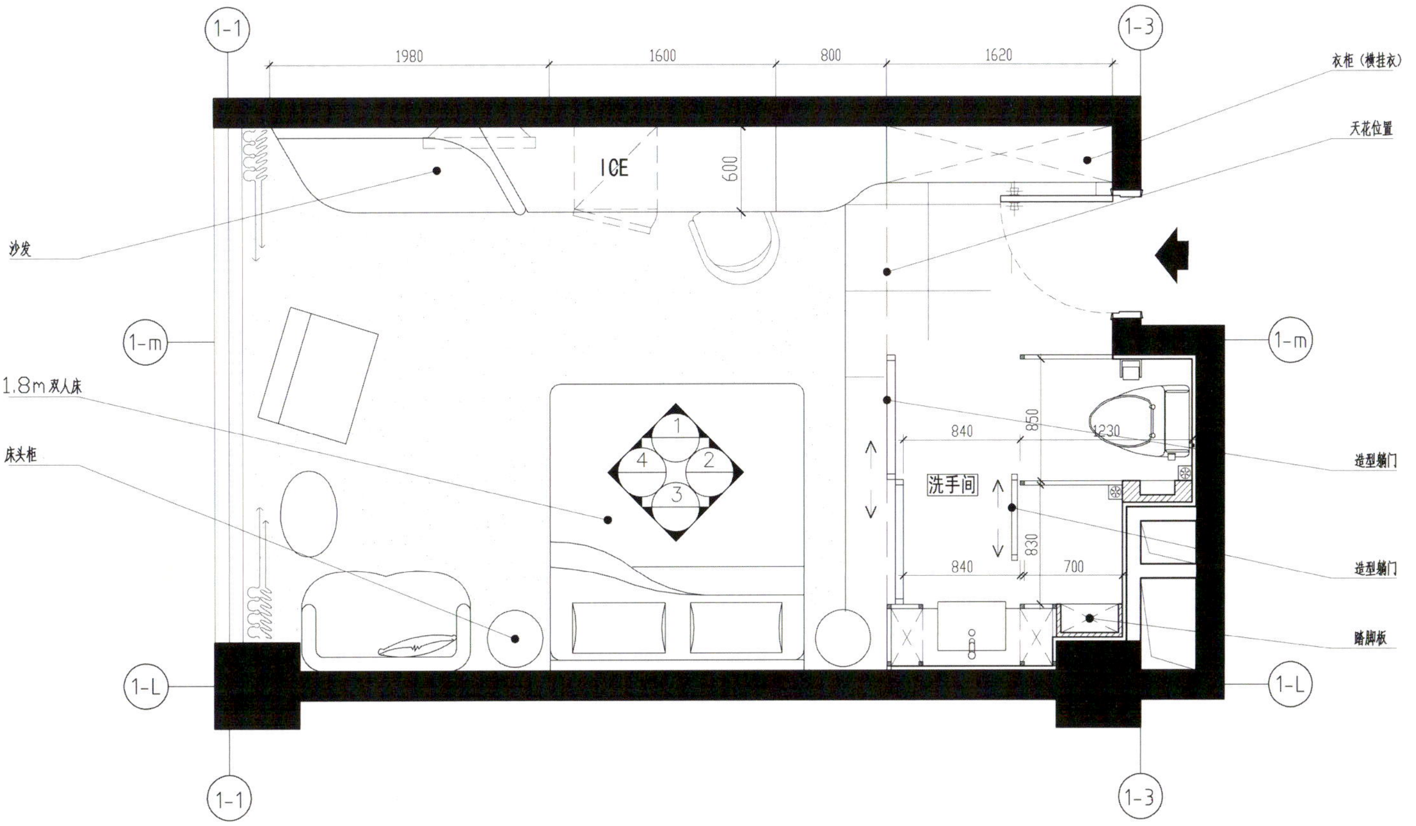

1-1
1-3
1980
1600
800
1620
衣柜（横挂衣）
天花位置
ICE
600
沙发
1-m
1.8m 双人床
床头柜
840
850
1230
洗手间
830
700
造型趟门
造型趟门
踏脚板
1-L

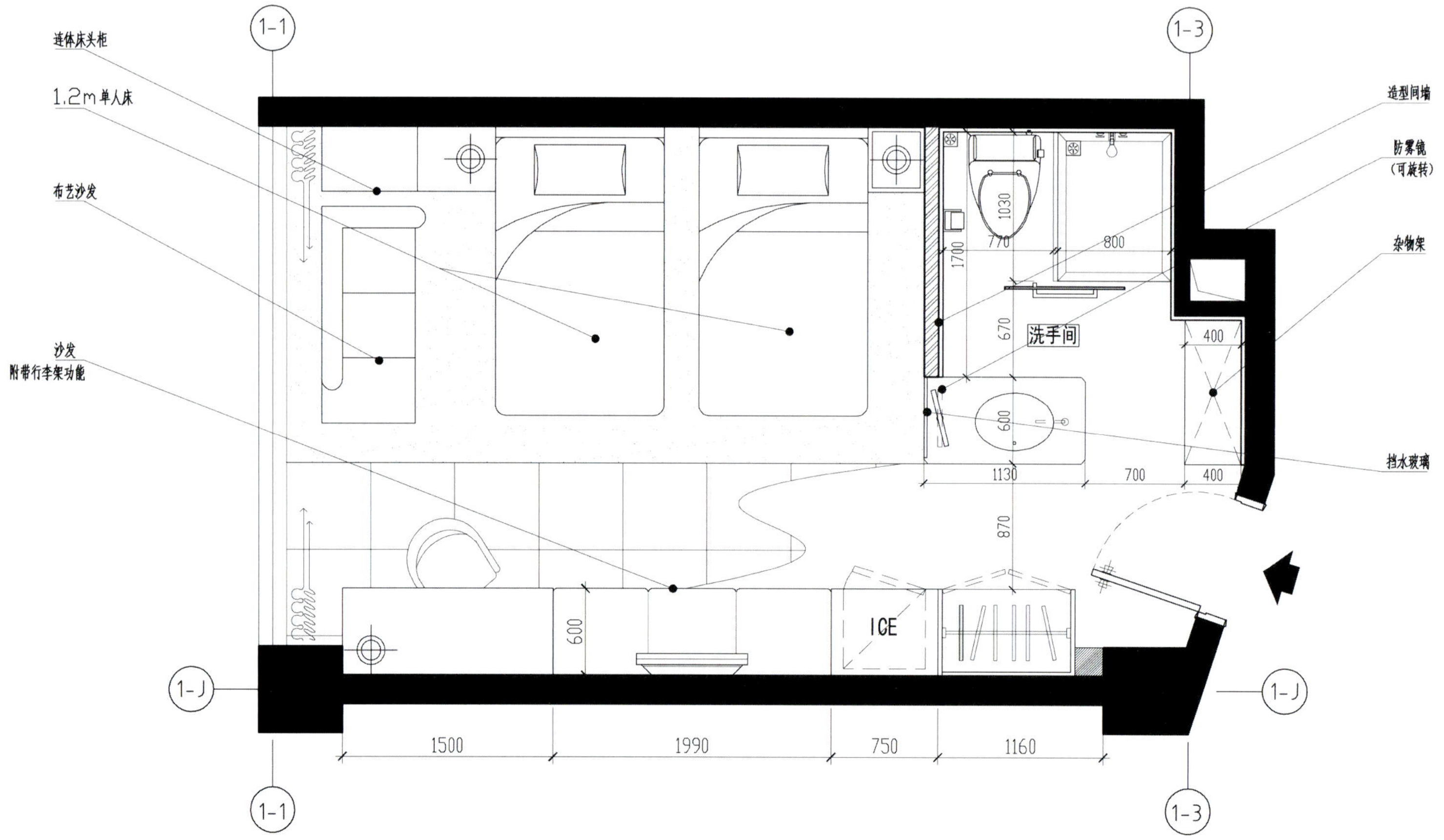

连体床头柜
1.2m 单人床
布艺沙发
沙发
附带行李架功能
造型间墙
防雾镜
(可旋转)
杂物架
挡水玻璃
洗手间
ICE
1-1
1-3
1-J
1030
770
800
1700
670
600
1130
700
400
870
1500
1990
750
1160

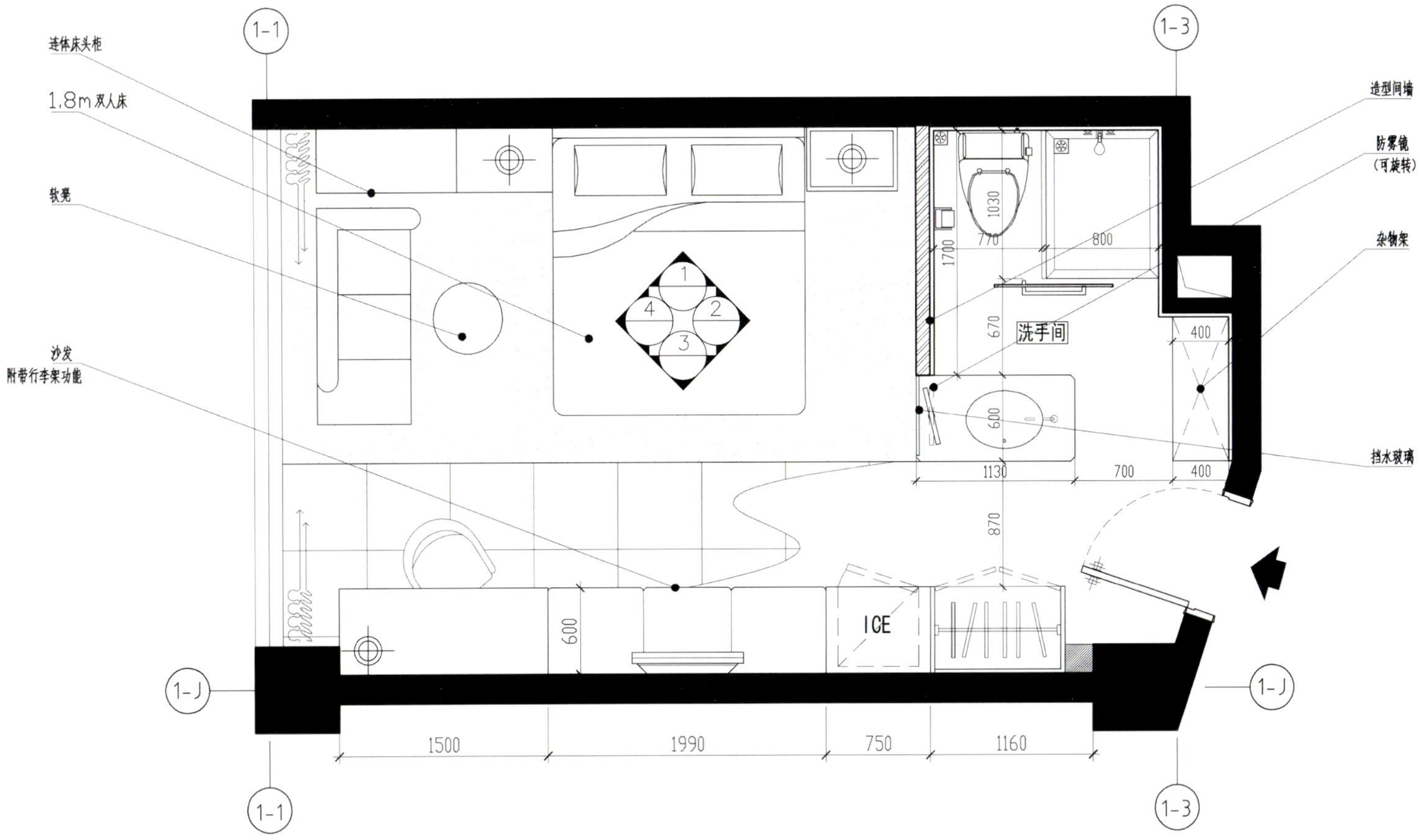

连体床头柜
1.8m 双人床
软凳
沙发
附带行李架功能
造型间墙
防雾镜
(可旋转)
杂物架
挡水玻璃
洗手间
ICE
1-1
1-3
1-J
1030
770
800
1700
670
400
600
1130
700
870
1500
1990
750
1160

地点	面积	设计师	设计公司	主要材料
/ 宜昌	/ 35000m²	/ 王治、范辉	/ 武汉市IEA设计顾问有限公司	/ 石材、地毯、墙纸、实木、玻璃、皮料

Ramada Yichang Hotel

华美达宜昌大酒店

Yichang, the hydropower capital in China, is well known for its beautiful cultural landscape and pleasant natural environment. In recent years, Yichang has become a tour destination in the central part of China. In this dynamic city that attracts the attention of the whole world, Ramada Hotel shows both the profound Chinese culture and the temperament matching with visitors from all over the world. The space is defined as grand and luxurious while giving full consideration to the privacy and honored experiences of the customer. Space divisions with rich Chinese philosophy and decorative language combining east with west convey both the profound cultural feeling and the relaxing life experience.

The hotel consists of a 5-storey podium, an old tower and a new one. The podium includes the hotel lobby, lounge, oriental food section, western food section, conference room and recreation services; and the towers include the standard guest rooms, suites, luxurious suites and apartments. By using different materials and lighting elements, the designers create a warm and delicate atmosphere and provide a truly unique accommodation experience.

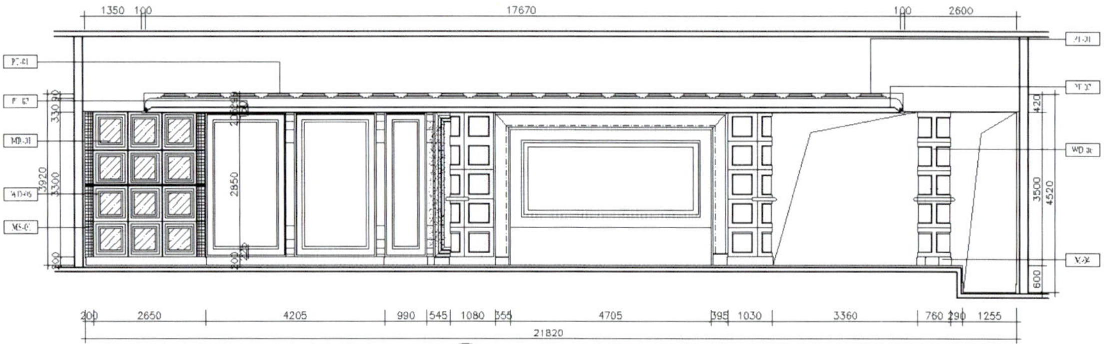

11 SS-1F-AL08
PART ELEVATION 立面图
一层大堂吧 scale 1:50

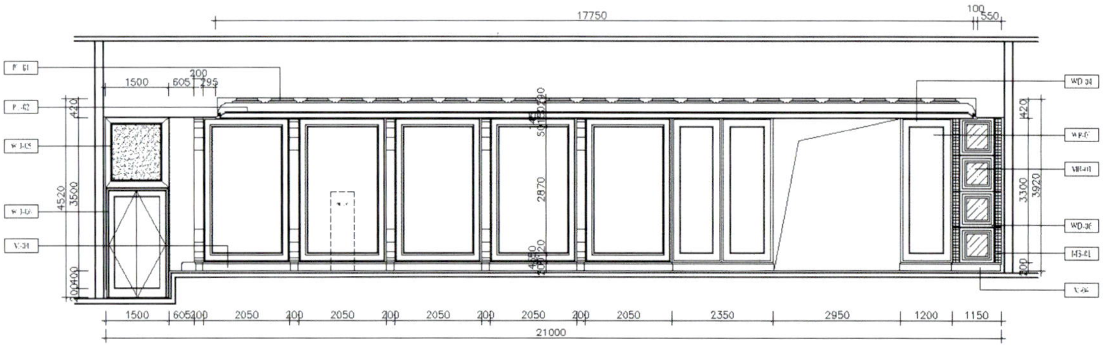

12 SS-1F-AL08
PART ELEVATION 立面图
一层大堂吧 scale 1:50

中国水电之都宜昌以优美的人文景观和怡人的自然环境著称于世。近年来，逐步成为中部旅游的目的地。在这样一个汇聚全球目光的活力之城，华美达大酒店要表现出中国文化的深厚意蕴及与世界各地来宾相匹配的气质。设计师将空间风格定义为隆重而奢华，充分考虑客人的隐私需求与尊荣体验，通过充满中式哲学意味的空间划分与融贯中西的装饰语言，传递出丰富的文化感受和轻松的生活体验。

酒店的整体建筑是由五层的裙楼和新旧两栋塔楼组成的。裙楼包括了酒店的大堂、酒廊、中餐厅、西餐厅、会议室和康乐设施；塔楼包括了酒店的标准客房、套房、豪华套房及公寓。设计师通过材料、照明等元素的运用，营造出温馨、精致的氛围，创造出真正独特的住宿体验。

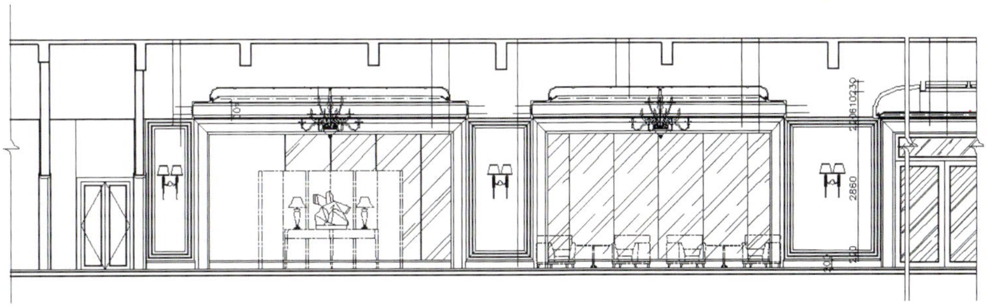

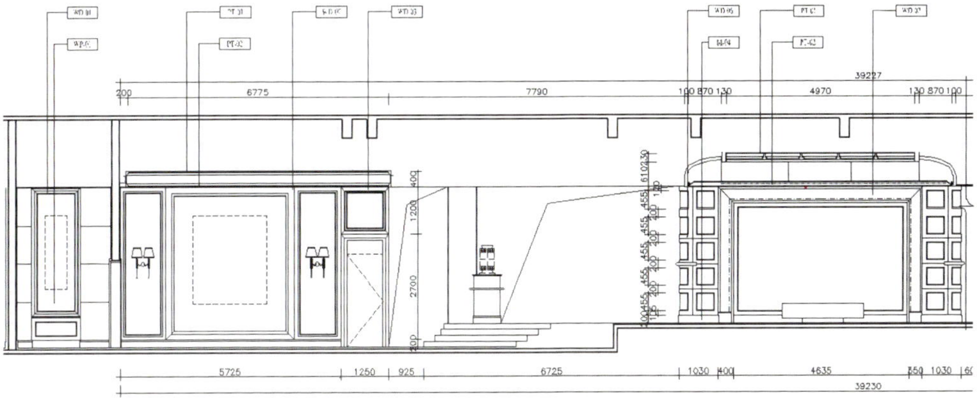

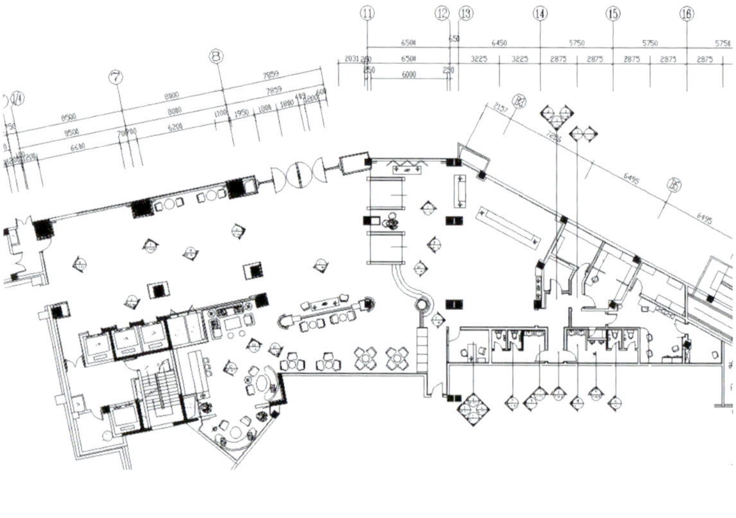

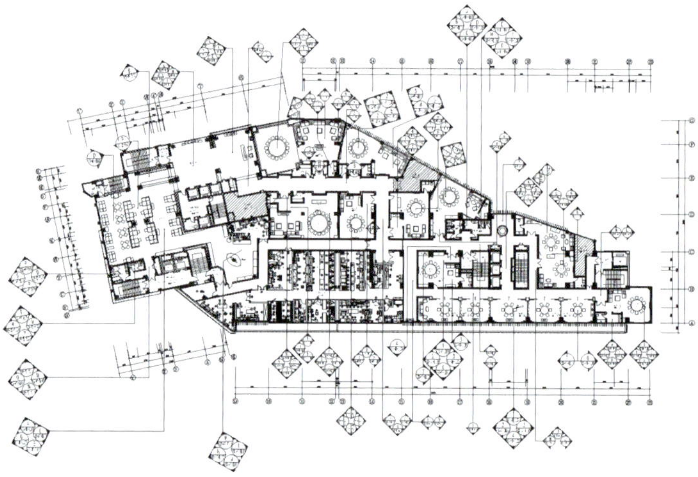

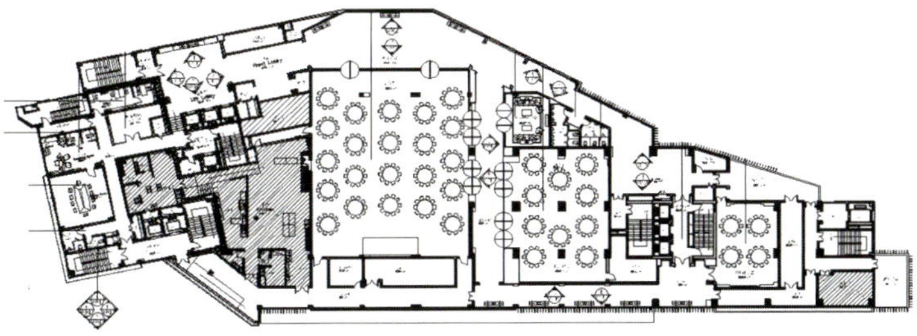

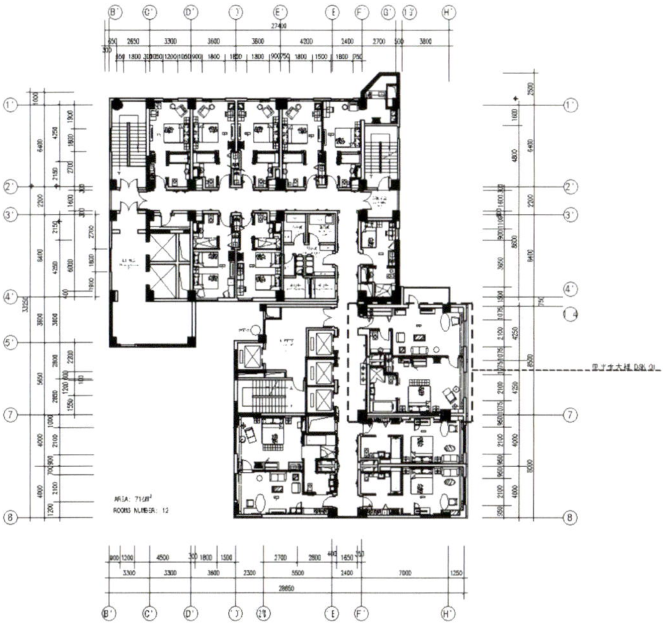

HKB 漢口銀行
热烈庆祝汉口银行宜昌分行
成立一周年

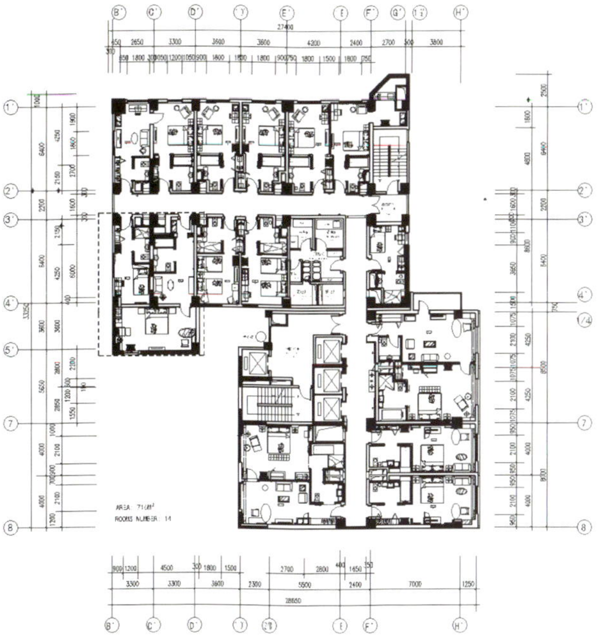

地点	面 积	设计师	设计公司	主要材料
/ 厦门	/ 1200m²	/ 李泷	/ 宽品设计顾问有限公司	/ 石材、木地板、涂料

Nazhai Boutique Hotel

那宅精品酒店

The designer faithfully preserves the historical appearance of the architectural structure and exteriror facade. With a selection of distinctive southern Fujian-styled washed stone for exterior walls, and plain, pure gray color as the tone, it blends with the local architectural culture and the features of Gulangyu. The public space is designed according to the concept of "Diamond Tower", with element of diamond cutting surface used in many aspects such as the spatial layout, facade modelling and the specific details. Taking the above as the thematic entry point, the design refines elements with rich humanistic caring and distinctive visual features, such as the customized nostalgic murals, the wooden floor puzzles, and the delicate-texture elevation materials. Adopting modern design concept and luxurious furnishing, the designer strives to create a fine-quality stylist atmosphere while echoing the overall design style, which brings resonances and delivers unique charm to the viewers and audiences.

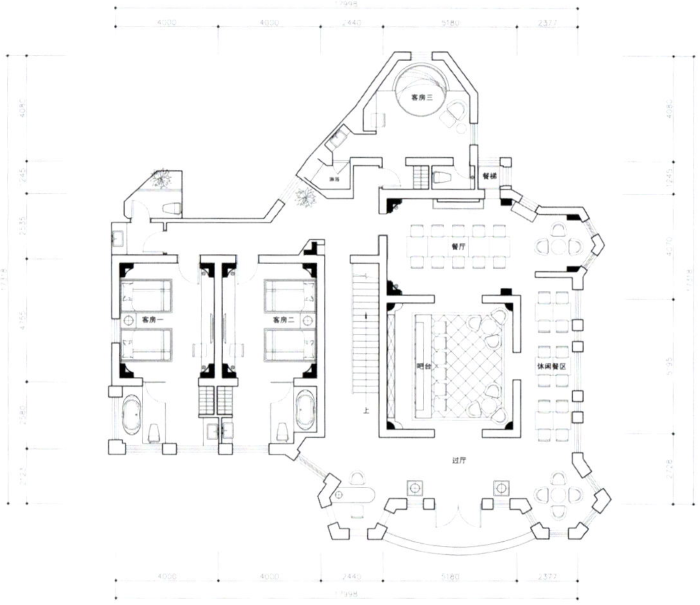

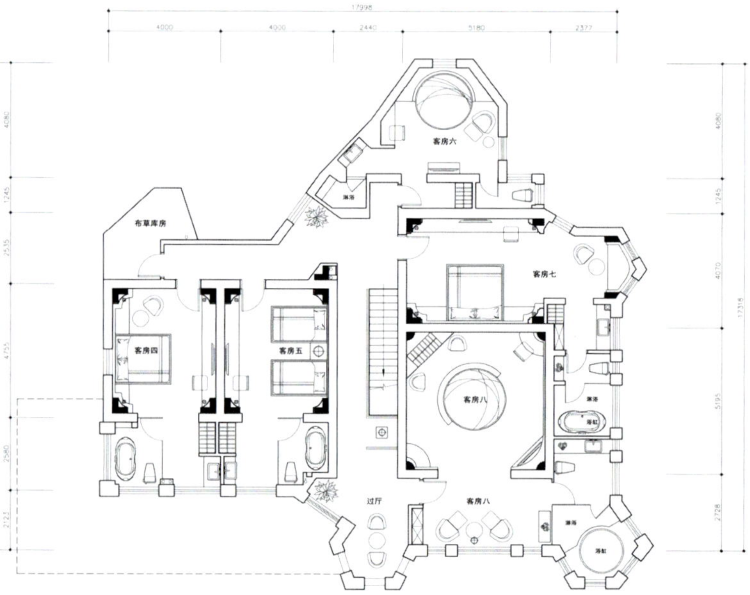

设计师让建筑结构和外观立面忠实保留历史原貌，选用具有鲜明闽南风格特征的水洗石外墙，配以质朴、纯净的浅灰色调，与当地特有的建筑文化环境及鼓浪屿特色融为一体。公共空间规划贯穿建筑“钻石楼”的设计理念，从空间布局、立面造型、局部细节等多处运用钻石切割面元素，设计更以此作为主题性的切入点，提炼具有丰富人文情怀及鲜明视觉特征的元素，如定制怀旧壁画、木地板拼图、具有细腻肌理的立面材质等，结合现代设计理念及奢华陈设，力求在呼应整体规划设计风格的同时，营造优良质感的时尚氛围，使观者及受众产生共鸣，感受优质空间的独特魅力。

地点	设计师	设计公司	摄影师
/ 伦敦	/ Jo Sampson, Tim Mutton	/ Blacksheep	/ Gareth Gardner

Novotel Tower Bridge Hotel

诺富特伦敦塔桥酒店

Novotel Tower Bridge Hotel is a four-star hotel in London, adjacent to the Tower of London and St. Paul's Cathedral and it was built in 1980s. The special geographic location is called "the flowing historyand" the interior architecture and design office Blacksheep meant to construct a destination of tourism and convention and a landmark building of cultural communication.

The hotel opens outwards in three directions, leading respectively to the Pepys Street, Savage Gardens and Cooper's Row. There are 203 guest rooms with Internet access and pay-movies; breakfast, lunch and supper are served in the theme restaurant and there is also a Pepys Bar; besides serving food and drink, the private dining room can be used for business meetings and informal business activities as well; there are rooms for sauna and Turkish bath in the fitness center. The hotel is equipped with seven flexible meeting rooms with audio-visual equipments and wireless Internet services.

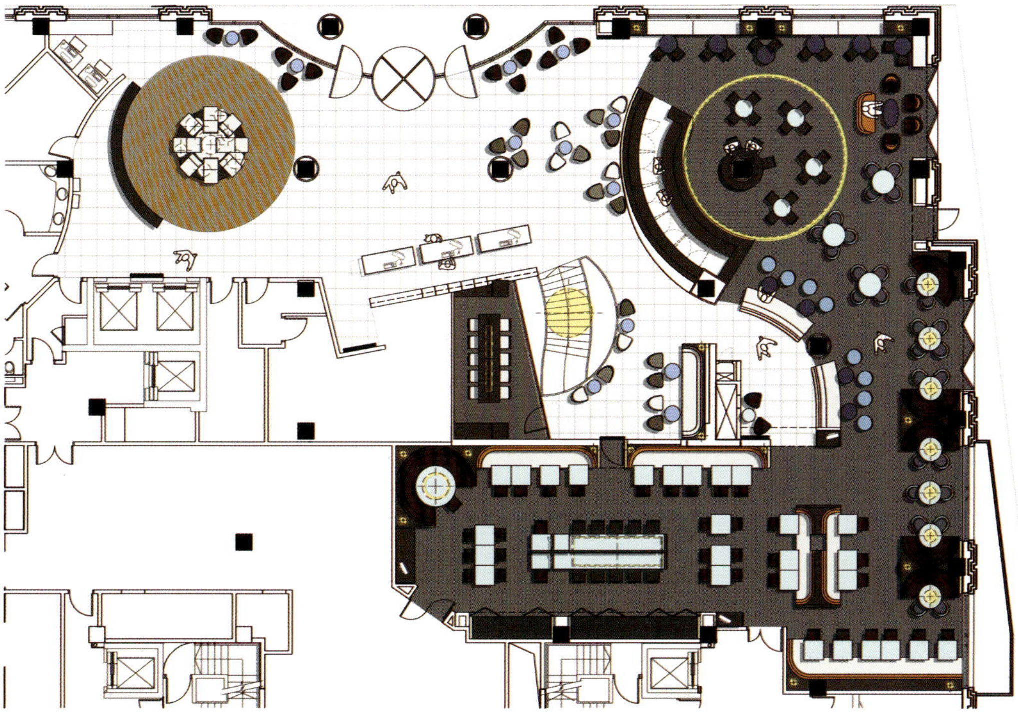

诺富特伦敦塔桥酒店是伦敦的一家四星级酒店。酒店位于伦敦市，邻近伦敦塔和圣保罗大教堂，始建于20世纪80年代。这个独特的地理位置被称为“流动的历史”，Blacksheep设计公司要在这里建造一个旅游集会的目的地及文化交流的标志性建筑。

酒店三面朝外，分别通往佩皮斯街、野人花园和古柏行。酒店内设203间客房，提供上网服务和付费电影；主题餐厅供应早、中、晚餐，还有Pepys酒吧；私人饭厅除了供应饮食，还可以用来作为商业会议和非正式的商务活动空间；健身中心有桑拿室和土耳其式浴室。酒店设有7间可灵活调整的会议室，里面音像设备齐全，并提供无线上网服务。

面积	设计公司	主要材料
/ 6000m²	/ ZEPPOS - GEORGIADI+ ASSOCIATES	/ 橡木地板、胡桃木地板、彩色玻璃、石膏板、马赛克等

Fresh Hotel

福来士酒店

The hotel is located in the downtown, a rich and colourful district with multinational populations. The original building was constructed in 1972 and it has been running as a hotel ever since. The renovation this time removed the beams, pillars and brick structures and the designer meant to communicate with both the environment and the people through the new building. On one hand, the building reflects the liveliness of the surrounding living environment–rich vignetting effect is created through different lighting on the bright colours; on the other hand, it is a new mark of fashion and has built this visible "modern fashion" by means of modern technology and materials. Painted glass, Corian and the advanced illumination make the hotel exist as a lighthouse, leading the trend of local characteristics; at the same time, as a decoration of its location, it adds to the beauty of the local environment.

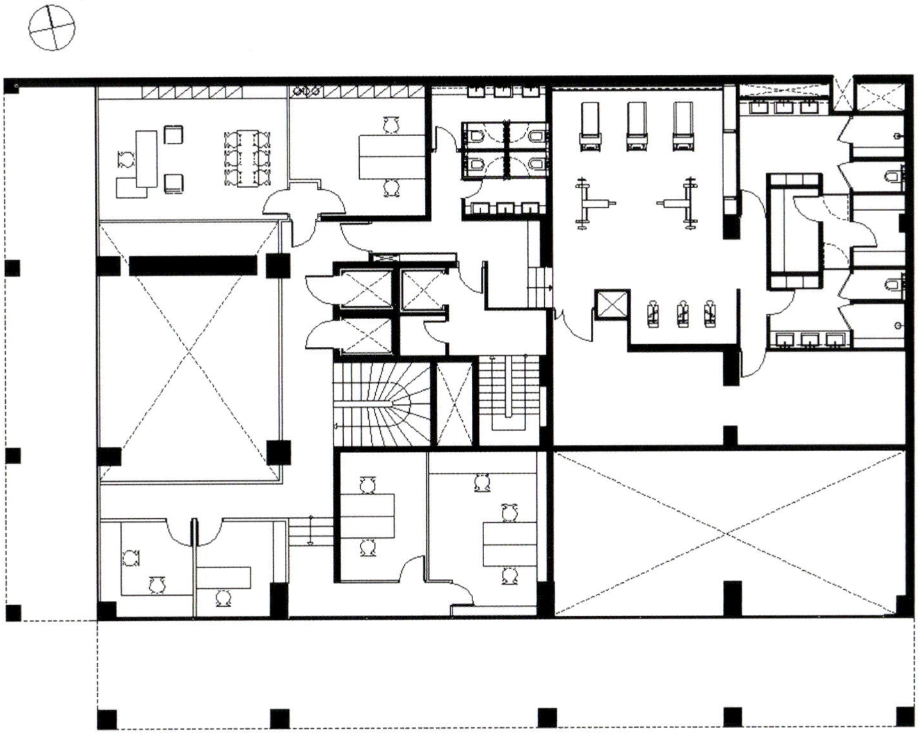

该店位于繁华的市中心，是一个丰富多彩的多民族地区。原建筑是1972年建造的，一直作为酒店来运营，本次整修拆掉了梁、柱和砖等结构，设计师要用一个新的建筑来与环境和人进行双重的对话。它一方面反映了周边生活环境的生动性——用鲜艳的色彩在不同照明条件下形成了丰富多彩的光晕效应；另一方面，它是一个新的时尚标，通过使用现代技术和材料创造了一个可视化的“现代时尚”。彩色玻璃、可丽耐和先进的照明方式让这个酒店像灯塔似的存在着，引领地方特色潮流；同时，作为地方上的一个点缀，也扮靓了区域环境。

地点	面积	设计师
/ 意大利	/ 1400m²	/ Guido Ciompi (Mr)

Hotel Centurion Palace

百夫长宫廷酒店

The aim of this project is to design a fine work of art on the Grand Canal of Venice, a hotel joining art and enjoyment together. Every single guest room or suite here has its own characteristics which makes the hotel special. The colors are mostly from the paintings by Giorgione, Tiziano, and Tiepolo, the famous European classical artists, therefore, the space reveals a quality of classical culture and art; combined with the vault in the bathroom, a similar design as St. Mark's Cathedral, making the hotel an incomparable display of artistic quality both in general and in detail.

Setting in the canal as a huge mirror, water becomes an indispensable theme in this project. The lively crystal pendent lamps, the wall paintings, the furniture, and even the large pendent lamp at the entry of the lobby, all have highlighted this theme.

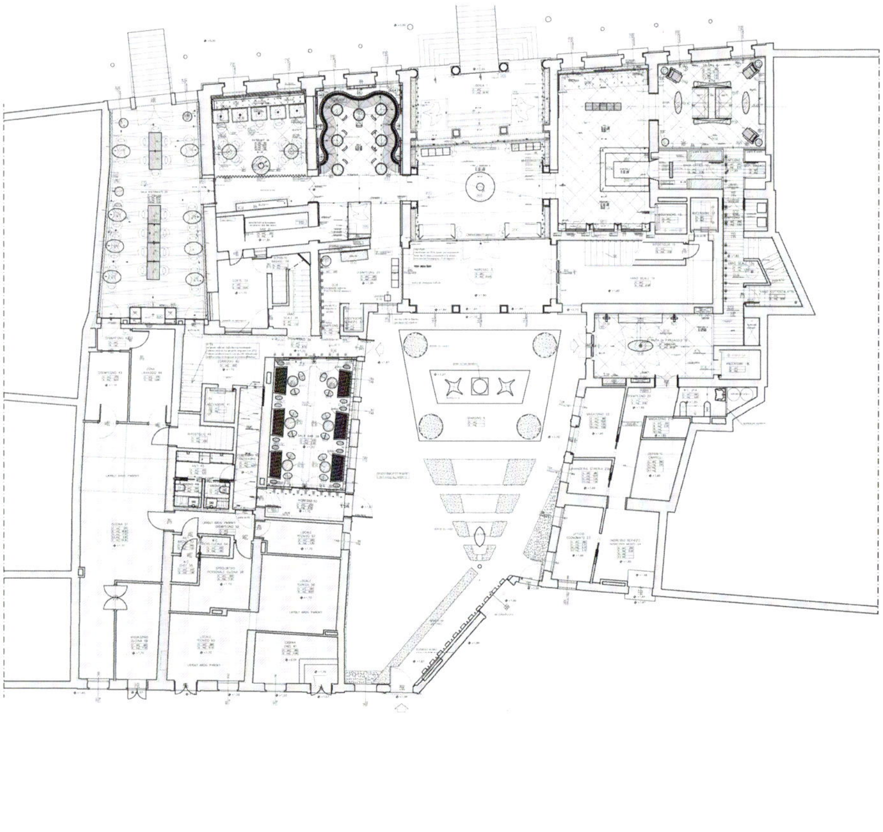

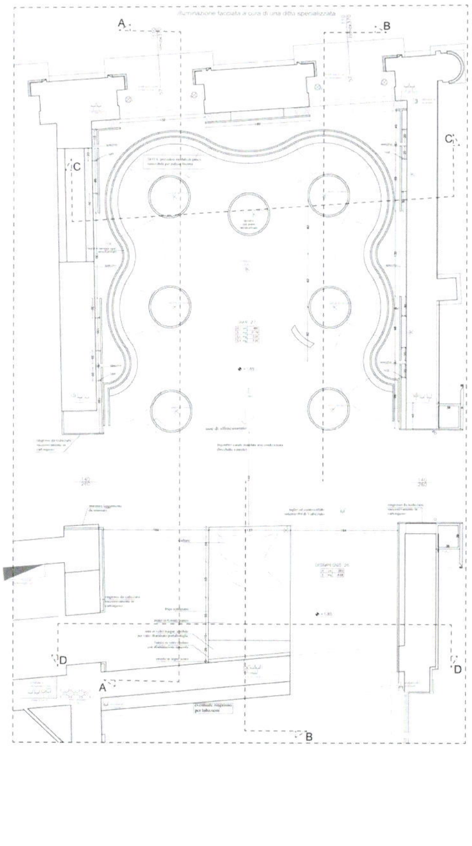

LG

本案的设计目标是要在威尼斯大运河上设计一个艺术精品，一个艺术与享受相结合的酒店。这里的每一间客房、套房都各有特色，酒店也因此显得独特。色彩方面，欧洲古典艺术名家乔尔乔内、提香和提埃波罗的画作，让空间不自觉地流露出古典文艺气质；加上浴室里与圣马可大教堂相似的拱顶设计，整个酒店无论是大局，还是细节，都展现出无与伦比的艺术气质。

酒店依托着大运河这面大镜子，因此，水这一主题在本案中也不可或缺。生动的水晶吊灯、壁画装饰、家具，甚至是大厅入口的大吊灯，都将这一主题凸显出来。

地点	面积	设计公司	主要材料
/ 广州市	/ 12112m²	/ 广州市韦格斯杨设计有限公司	/ 石材、木饰面、乳胶漆、瓷砖、马赛克

Nansha Haidecheng Hotel

南沙海的城酒店

Haidecheng Hotel is developed under Crystal Bay Project, the third phase of Nansha Coastal Garden project conducted by Urban Construction Group. Separated by a winding garden corridor, it is next to the Crystal Bay Villas in the style of small European towns. The building takes the shape of a long east-west strip, and the facade imitats the simple and refreshing Mediterranean tone. This forms a sharp contrast with the lushness around, but without causing any feeling of abruptness.

The hotel consists of four floors: on the first floor, there are the lobby, hall bar, buffet restaurant, and business conference area. On the second, third and fourth floors, there are standard rooms, household rooms, and suites, more than 90 in all. In the interior design, the external Mediterranean architectural appearance is reused and extended to the interior space. On the basis of this, the designer adds and enlarges some symbolic elements, repeats them in use, and makes the interior space abundant in its uniqueness, thus connecting different functional spaces in form, and making the whole architecture more unified, without any feeling of trivialness.

480

海的城酒店属于城建集团开发的南沙滨海花园三期“水晶湾”项目，与充满浓郁欧洲小镇风格的水晶湾别墅群只隔一道蜿蜒的花园走廊。长条形的建筑体呈东西走向，建筑外观风格则模仿了地中海简洁清爽的调子，与周围的葱郁形成了鲜明的对比，但又不会显得突兀。

酒店分四层：首层分布有大堂、堂吧、自助餐厅及商务会议区域。二、三、四层共分布标准房、家庭房、套房等共计九十余间。在室内设计中，设计师将建筑外观的地中海风格沿用、延伸至室内，在此基础上加入及放大了一些符号性的元素，并将之重复再重复，使室内空间充满了自身的特色，并使不同功能空间之间有了形式上的相连，令整体更为统一，但又不会令人感到繁琐。

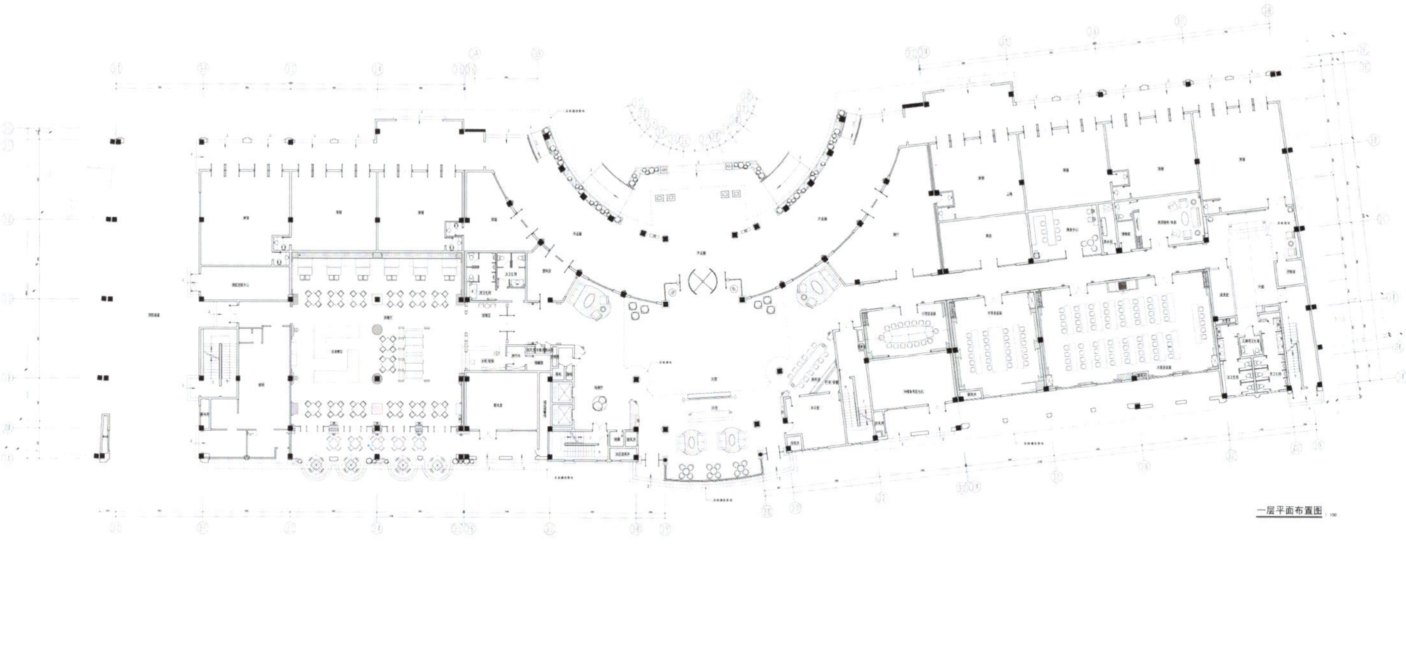

一层平面布置图

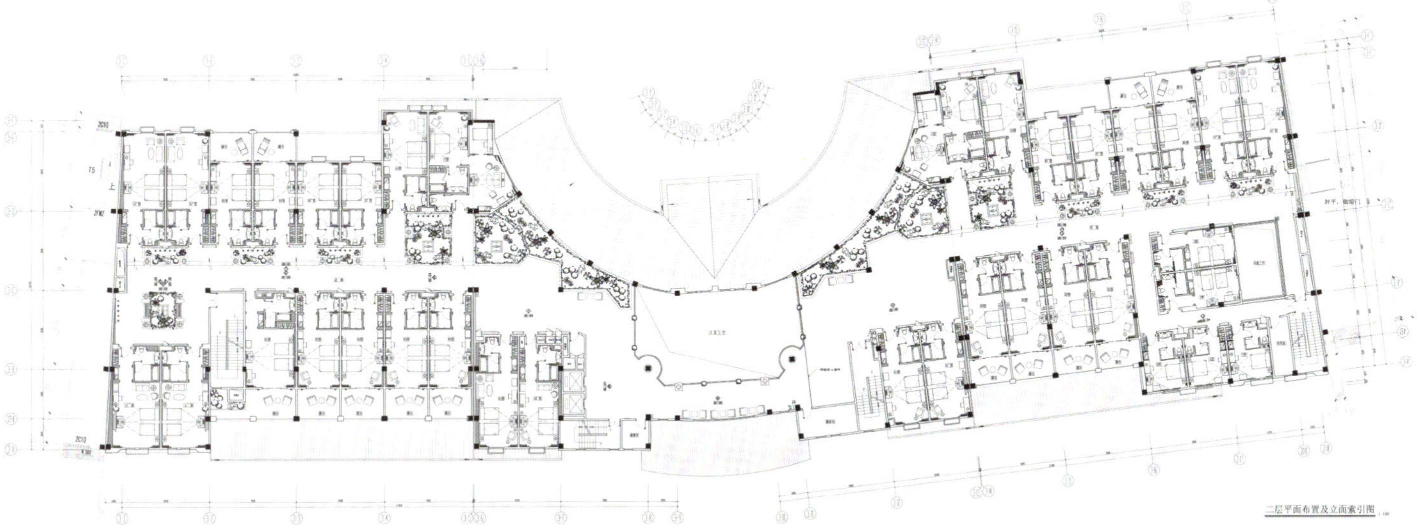

二层平面布置及立面索引图

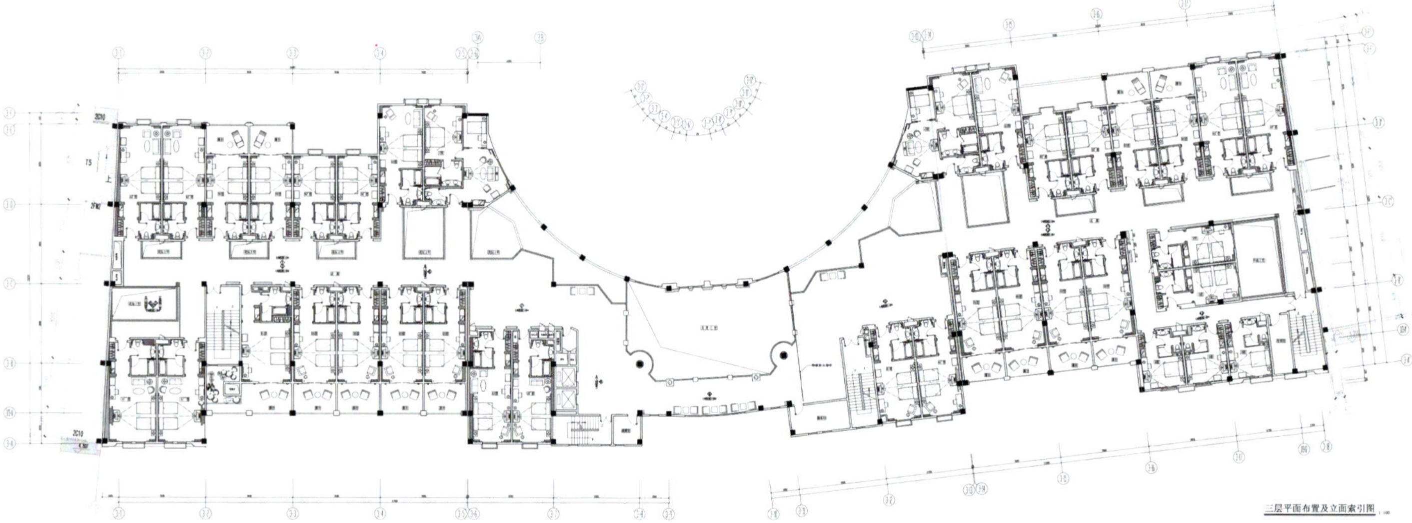

三层平面布置及立面索引图 1:100

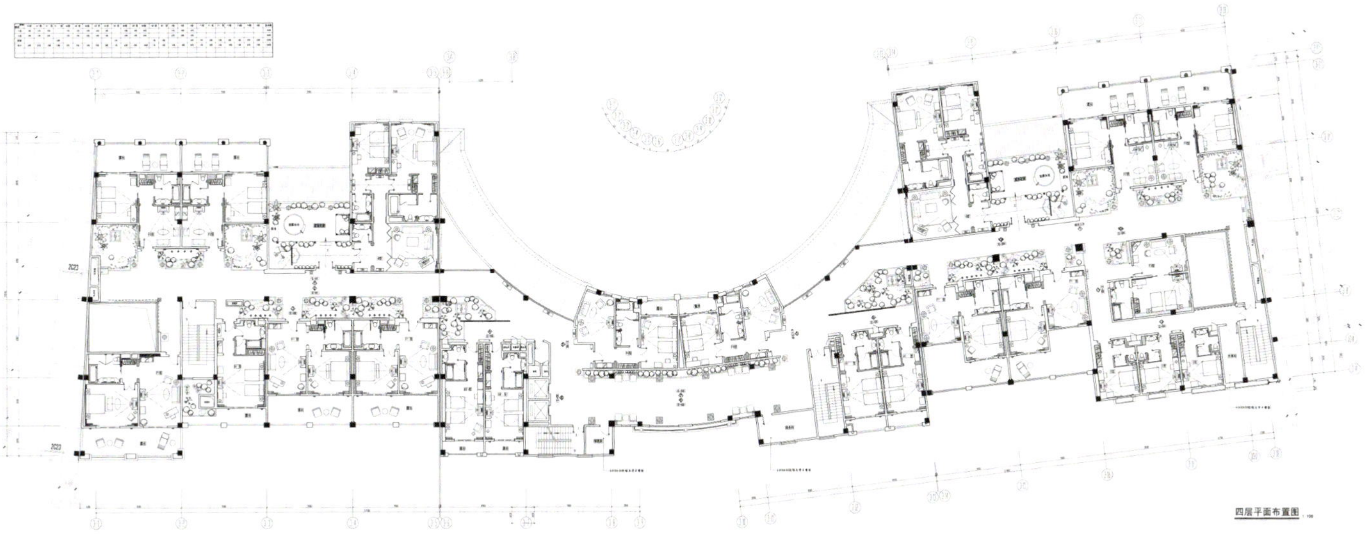

四层平面布置图

地点	面 积	设计师	陈设选材	设计公司	主要材料
/ 肇庆	/ 7000m²	/ 黄炽烽、欧敏华	/ 卢颖仪、谭启开	/ J2-DESIGN顾问设计有限公司	/ 白木纹、白橡木、防火板、陶艺等

Zhaoqing M Hotel

肇庆万仕酒店

This project chooses materials such as white wood grain, white oak, fireproof boards to create a modern, artistic space with characteristics. The LOGO at the hotel entrance looks eye-catching under the illumination and the letters of LOGO in the glass windows add some fashionable interests to the space. Entering the hotel lobby, the running water scenery on both sides of the anteroom brings life to the static scenery and it appears quieter and easier under the illumination. As the highlight of the hotel, ceramics combination on the background wall of the lobby reception fills the space with dynamic feelings. This design has built a fundamental atmosphere of fashion, quietness and leisure. When it extends to the guest rooms, the massively used materials of solid wood and the concise design make the space look cozy, quiet and elegant, which brings guests kind of relaxed mood.

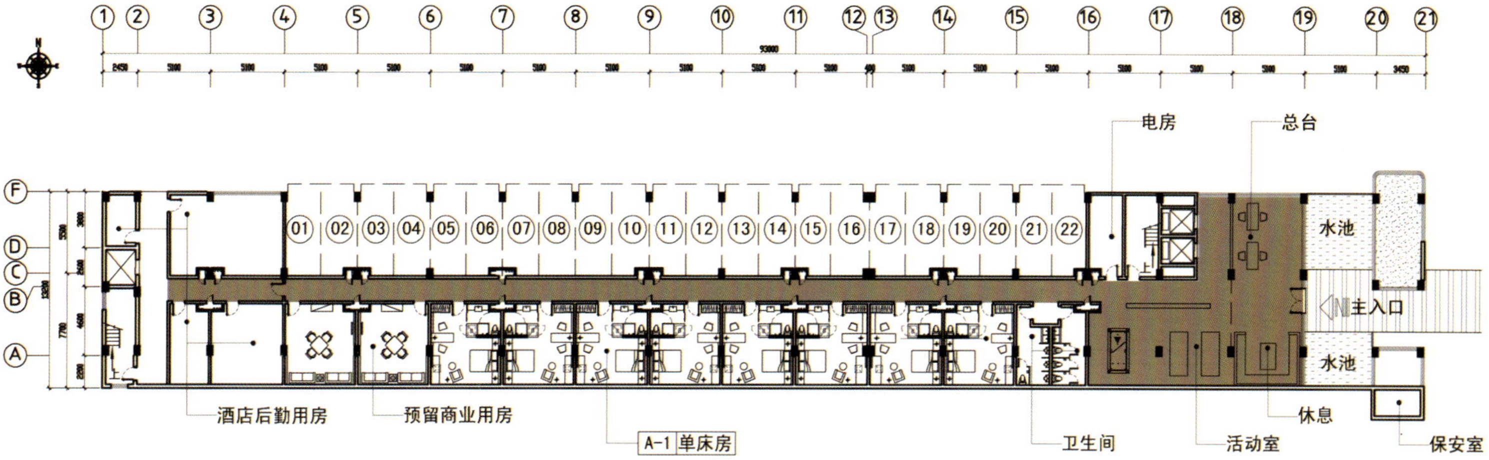

1
2
3
4
5
6
7
8
9
10
11
12
13
14
15
16
17
18
19
20
21
F
D
C
B
A
01
02
03
04
05
06
07
08
09
10
11
12
13
14
15
16
17
18
19
20
21
22
电房
总台
水池
主入口
水池
酒店后勤用房
预留商业用房
A-1 单床房
卫生间
活动室
休息
保安室

MHOTEL
萬仕酒店
Dugout Books

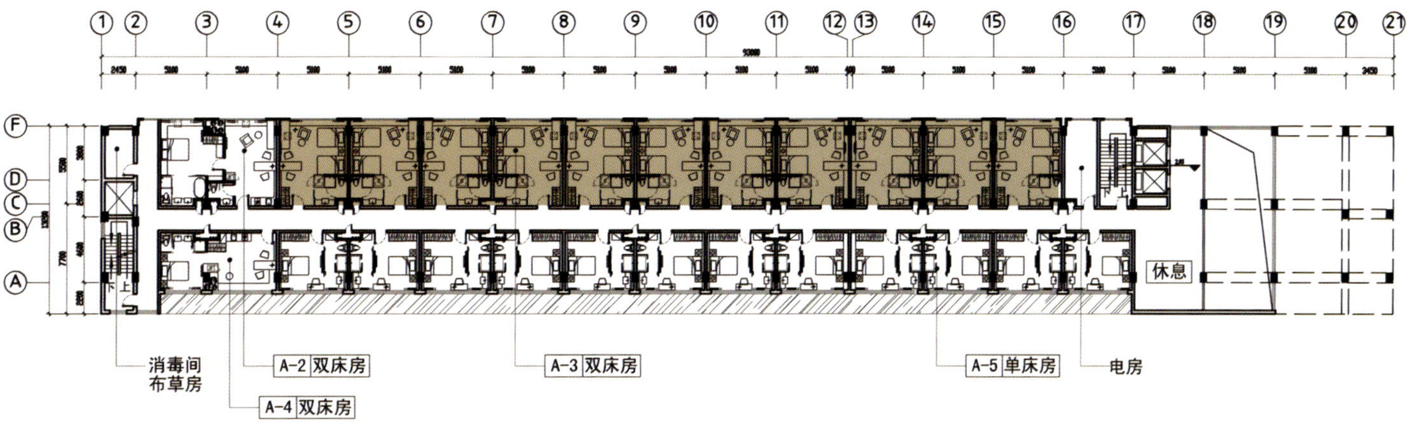
休息
消毒间
布草房
A-2 双床房
A-4 双床房
A-3 双床房
A-5 单床房
电房

M HOTEL
萬仕酒店
Dugout Books

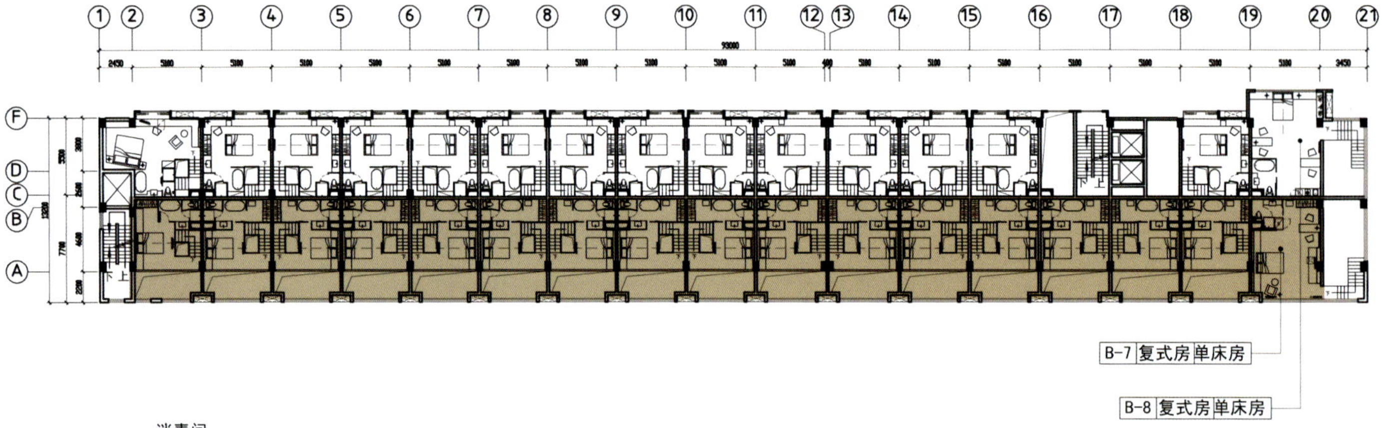
B-7 复式房单床房
B-8 复式房单床房

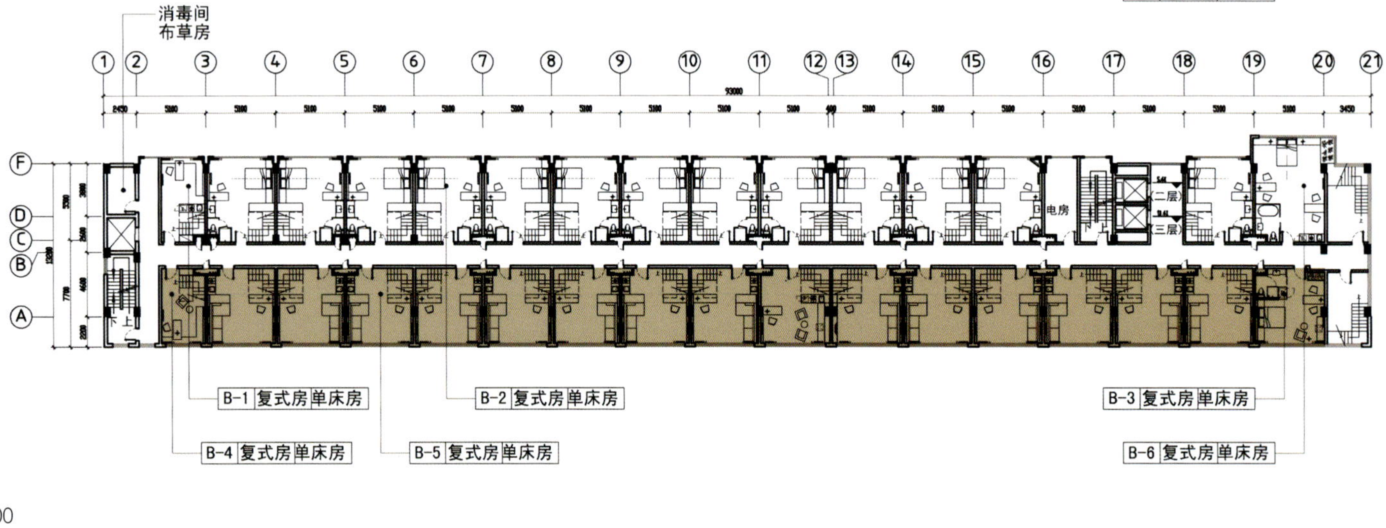
消毒间
布草房
电房
二层
三层
B-1 复式房单床房
B-2 复式房单床房
B-3 复式房单床房
B-4 复式房单床房
B-5 复式房单床房
B-6 复式房单床房

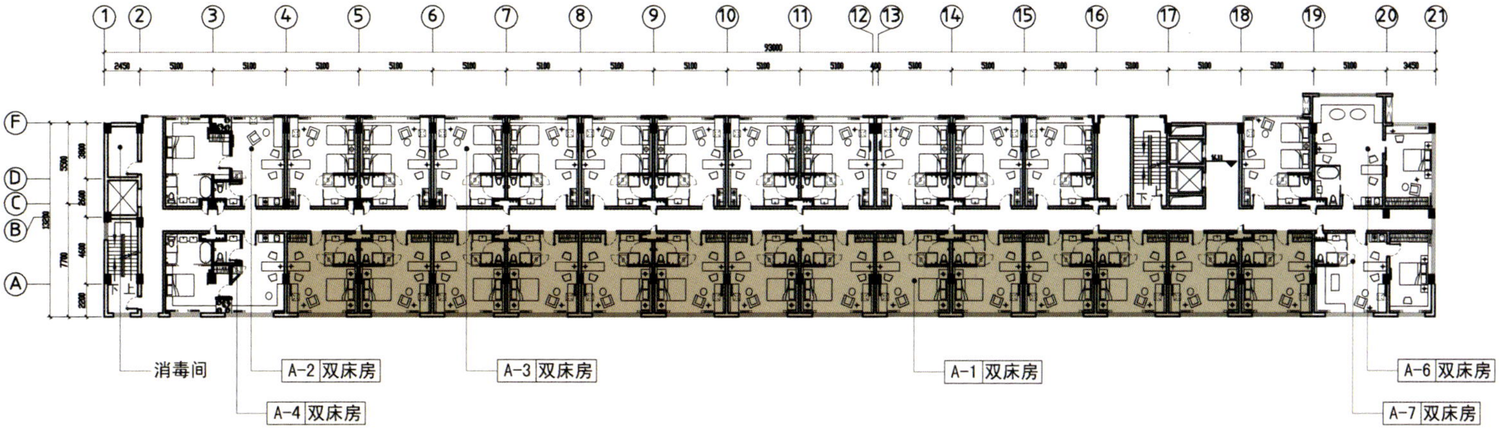

本案酒店选用白木纹、白橡木、防火板等材料，营造出现代、艺术、个性的空间。酒店入口处的标志在灯光的照射下显得格外醒目，玻璃窗内的标志字母为空间平添了几分时尚意趣。进入酒店大堂，前厅两边的流动水景让静止的风景鲜活起来，在灯光的映照下显得格外静逸。大堂接待背景墙的陶艺组合作为酒店的亮点，使空间充满动感。这样的设计已经为酒店奠定好一个时尚、静逸的基础氛围，延伸到客房里，实木材料的大量运用及简约的设计，让空间显得温馨静雅，带给人一种放松的心境。

地点	面积	设计师	设计公司	主要材料
/ 绍兴	/ 10000m²	/ 黄炽烽、欧敏华	/ J2-DESIGN顾问设计有限公司	/ 古木纹、灰木纹、胡桃木

Smart Hotel (Shaoxing)

时尚旅酒店(绍兴店)

In this project, the hotel lobby employs a large amount of stone to build the space. The floor employs antique wood grain and the walls employ gray wood grain side putting together with plane surface which creates the effect of abundant changes. In addition to the effect of lighting, it produces effects of light reflection like mirrors, which lightens the space. In the function of floor plan, a small-sized American billiards table is placed at the entrance of the lobby, which connects with the lobby bar, lounge and reception area to the side as one body. Dark hues are employed in the guest rooms to create mysterious atmosphere and the carpet audaciously adopts blue water wave color to set off embroidery hard board with water wave striate on the back of the bed, which adds somewhat of fashion and romance. The bathroom adopts the open style by employing materials such as steel glass to exemplify the modern and fashionable atmosphere in the entire guest room.

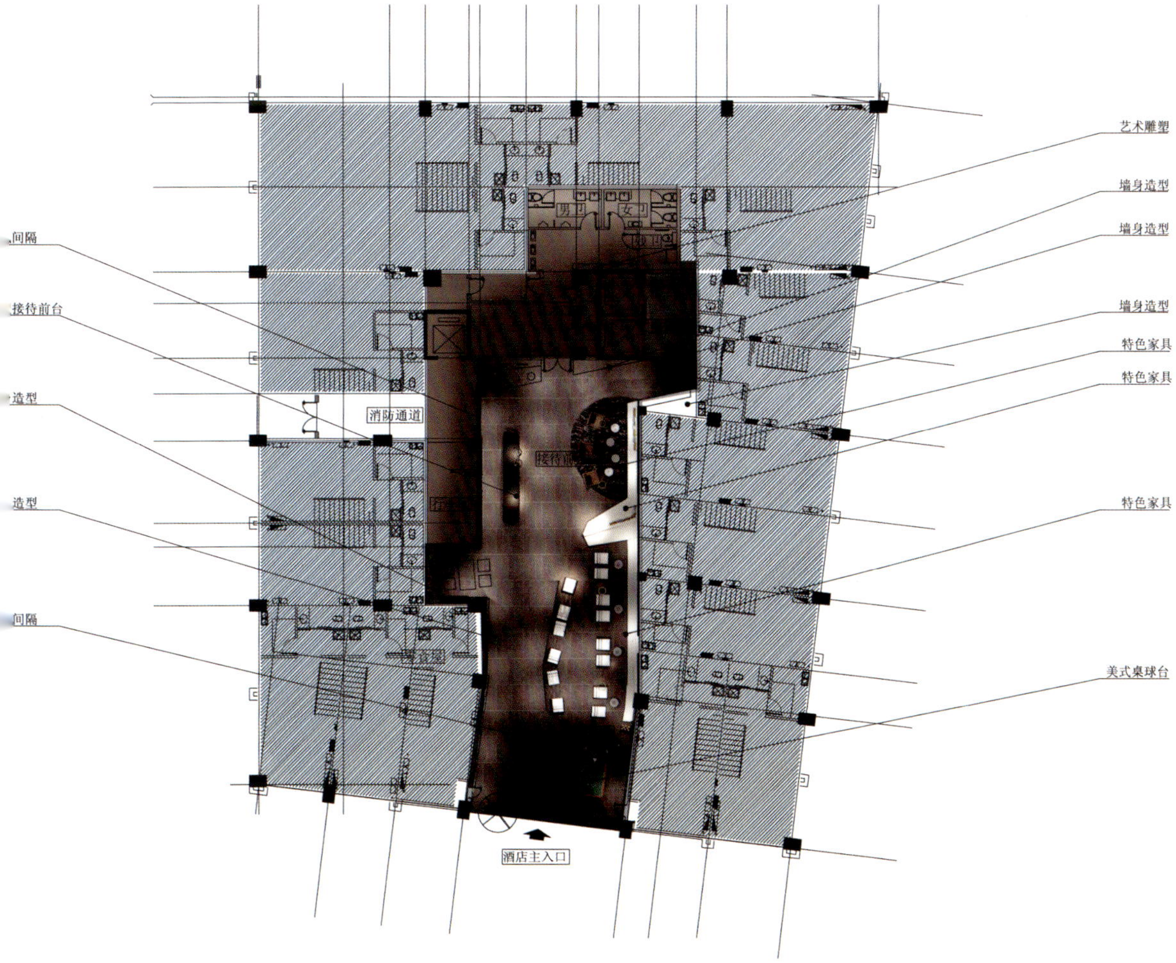

本案中，酒店大堂运用了大量的石材来打造空间，地面运用了古木纹，墙身运用了灰木纹光面错拼拉丝面，达到了丰富的变化效果。加上灯光的效果，产生一种镜面般的反光效应，让空间熠熠生辉。平面功能上，大堂入口设计了小型美式台球桌，与旁边大堂吧和休息接待处连成一体。客房运用深色调营造出神秘的气氛，地毯大胆地采用了水纹蓝色色彩，映衬着床背水纹刺绣皮纹硬板，更添几分时尚、浪漫。卫生间采用开放式处理，运用精钢玻璃等材料，体现出整个客房的现代时尚气息。

Bird's Nest Resort

鸟巢度假村

This project is located in China's only tropical coastal city: the Yalongwan National Resort of Sanya City, Hainan Province. With forest ecological environment as its main body, the park combines mountain landscape, biological landscape, cultural landscape and astrological landscape together. Characterized by ecological tour and recreation and being a multi-functional integrated forest park, it offers services such as leisure vacation, entertainment fitness and ecological protection education.

Among the original tropical rain forest, staging tents are set up for the customers to enjoy mountain vacation and natural ecology. With more than 20 semi-open tents built alongside the mountains, guests will have the feeling of returning to nature and approaching nature. Such natural materials as wood, bark, and grass are used as main elements for the architectural design of the Bird's nest. While overlooking the sea, it is built alongside the mountain, with nature and ecology adding beauty to each other. The building is characterized by the tropical atmosphere, plain but modestly extravagant. Dwelling above the forest with clouds around, and overlooking the sea joining with the sky, visitors will enjoy thorough relaxation both physically and mentally.

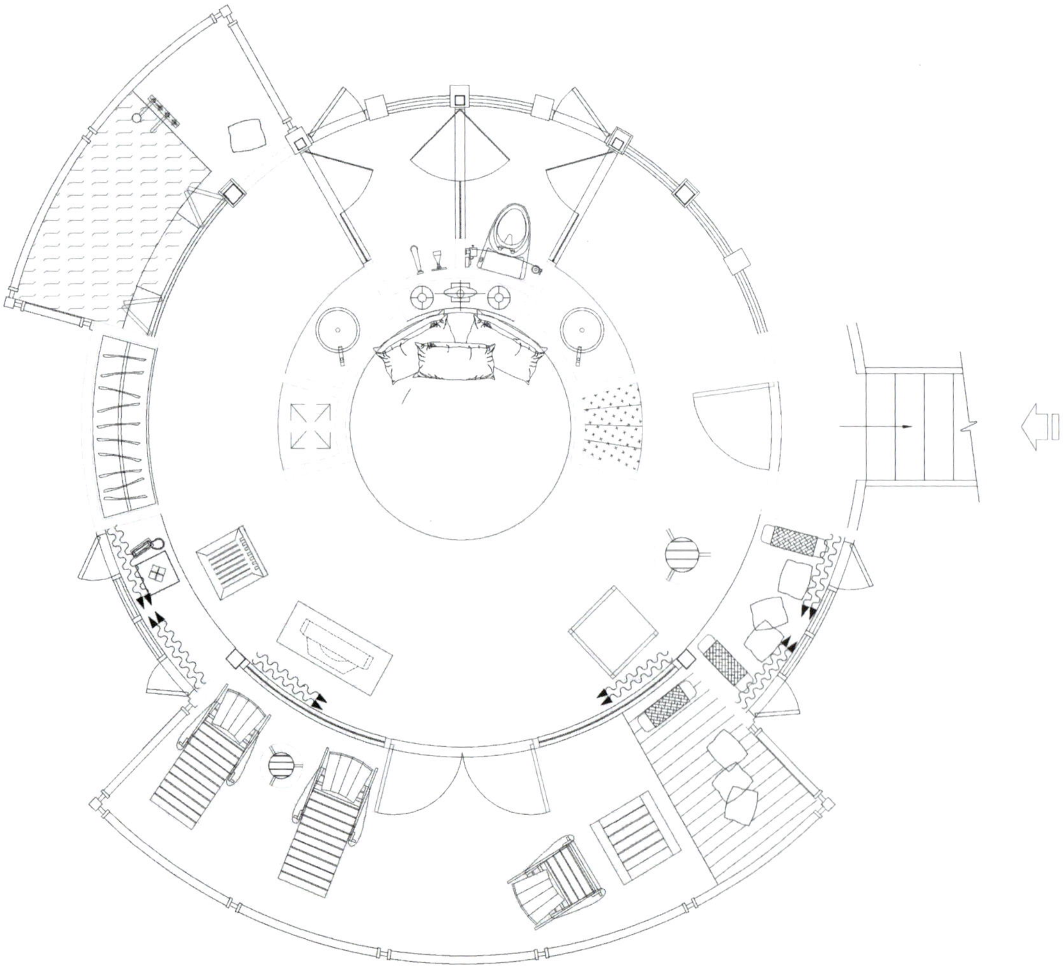

项目位于中国唯一的热带滨海城市——海南省三亚市的亚龙湾国家旅游度假区。公园以森林生态环境为主体，集山体景观、生物景观、人文景观、天象景观于一体，是以生态游憩为特色，融休闲度假、娱乐健身、生态保护教育等为主要内容的多功能综合性森林公园。

在原生态的热带雨林间设计了以山地度假休闲和自然生态为主的集结地帐篷房，20多间半敞开式的帐篷房依山而建，为客人营造一种回归自然、亲近自然的氛围。鸟巢度假村在建筑设计上以木材、树皮、茅草等原生态的天然建材为主要元素，伴山面海而建，自然与生态交相辉映。建筑独具热带风情，质朴之内又显低调奢华，栖居于丛林之上，云雾袅袅，远眺海天一线，使游客全身心放松。

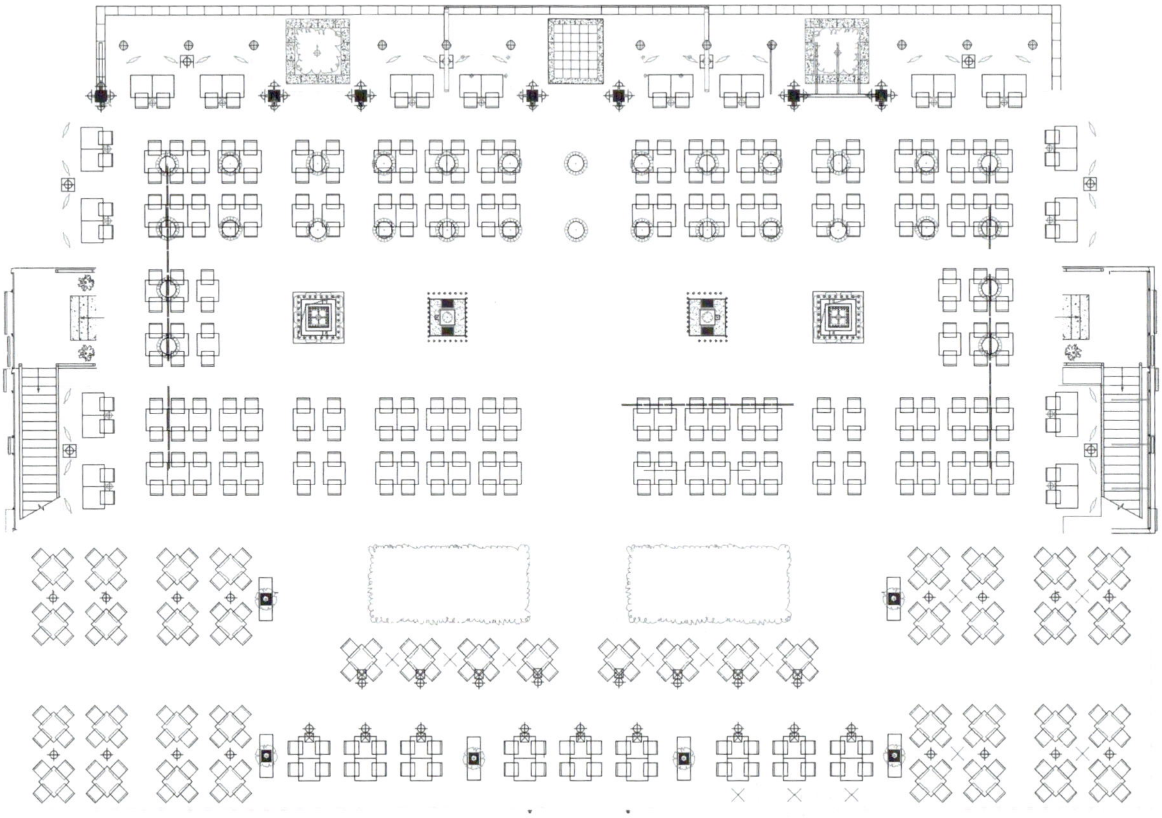

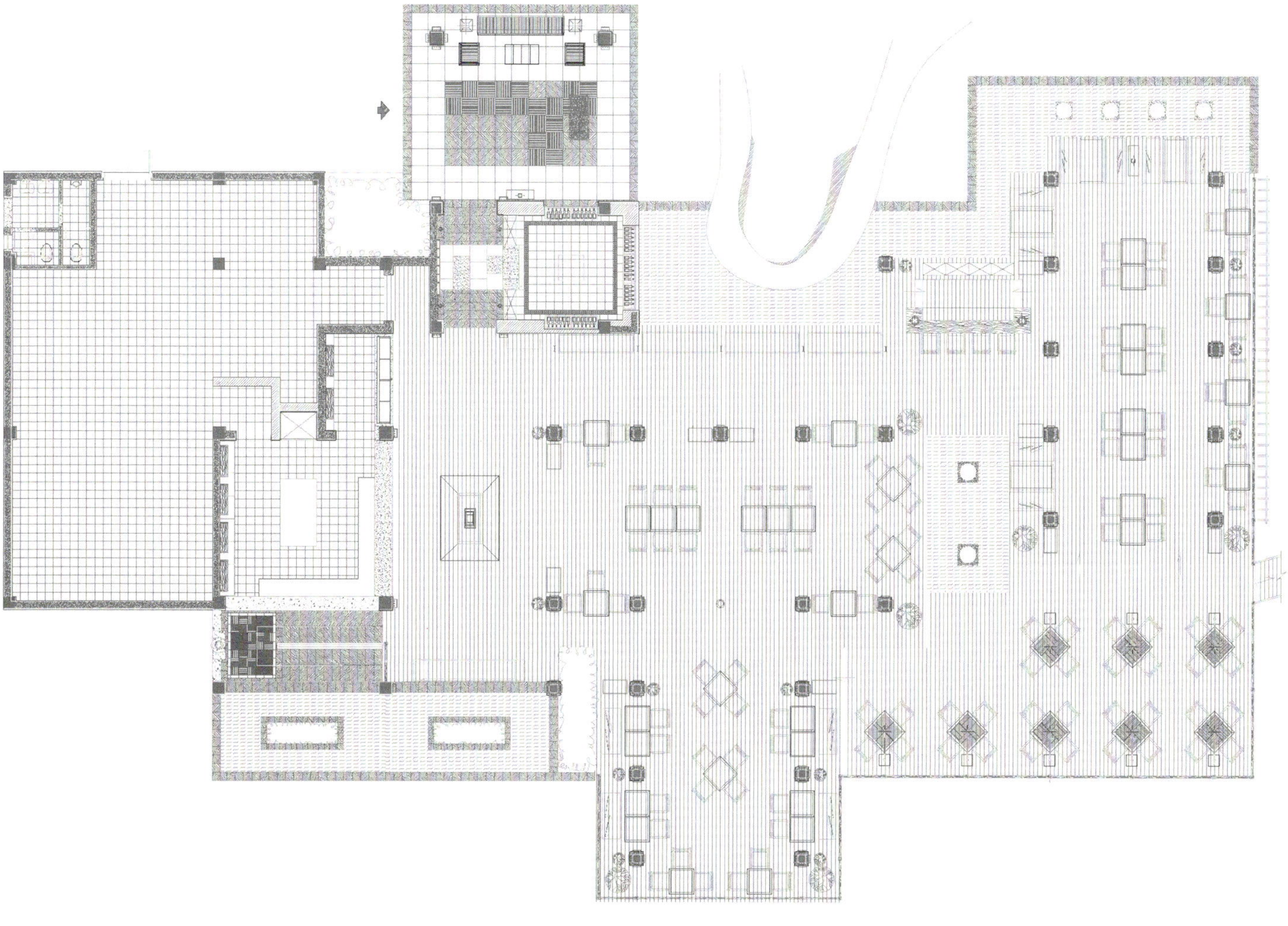

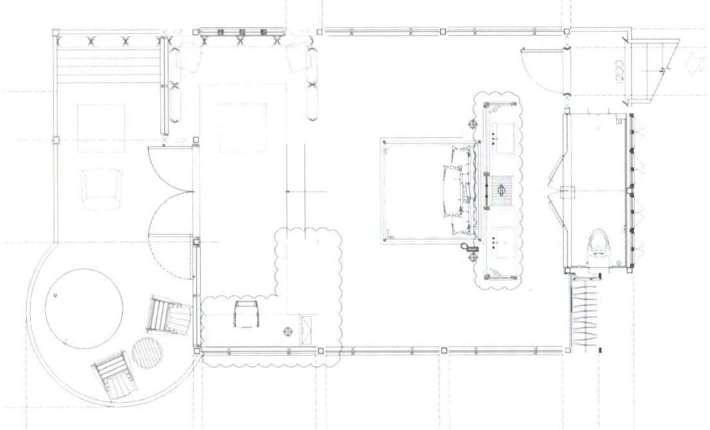

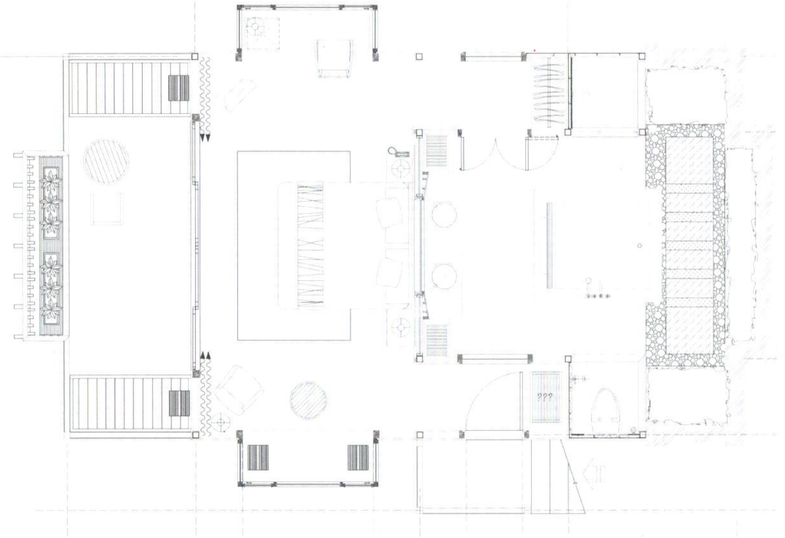

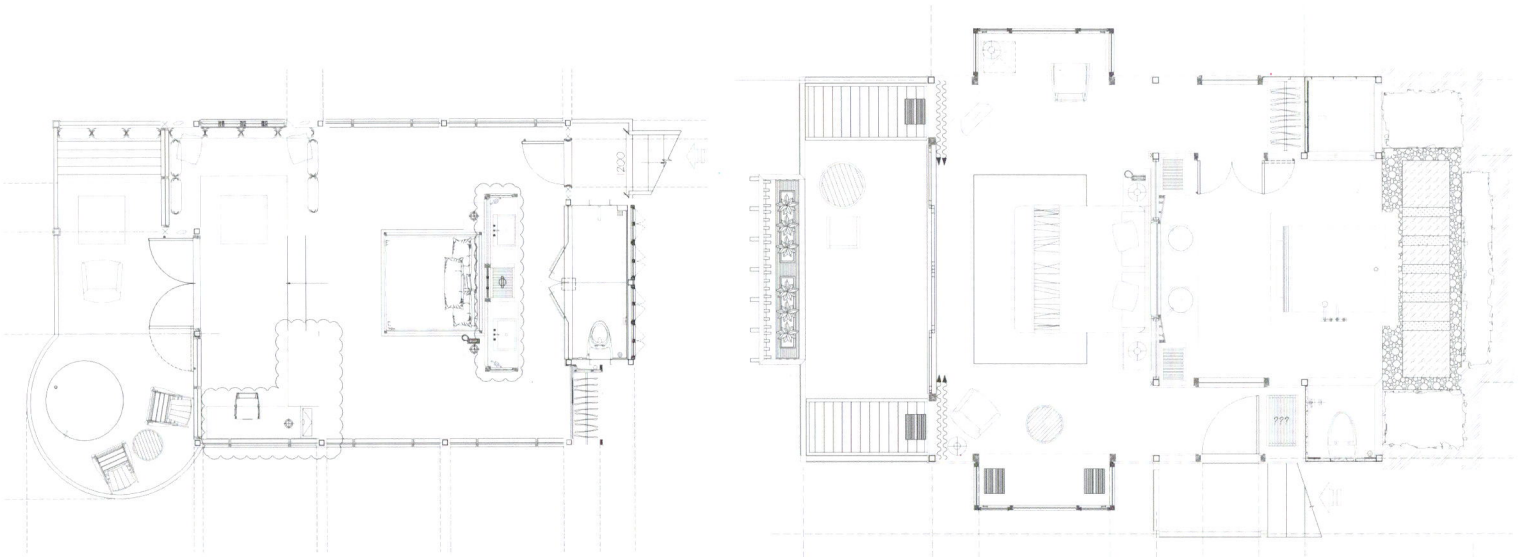

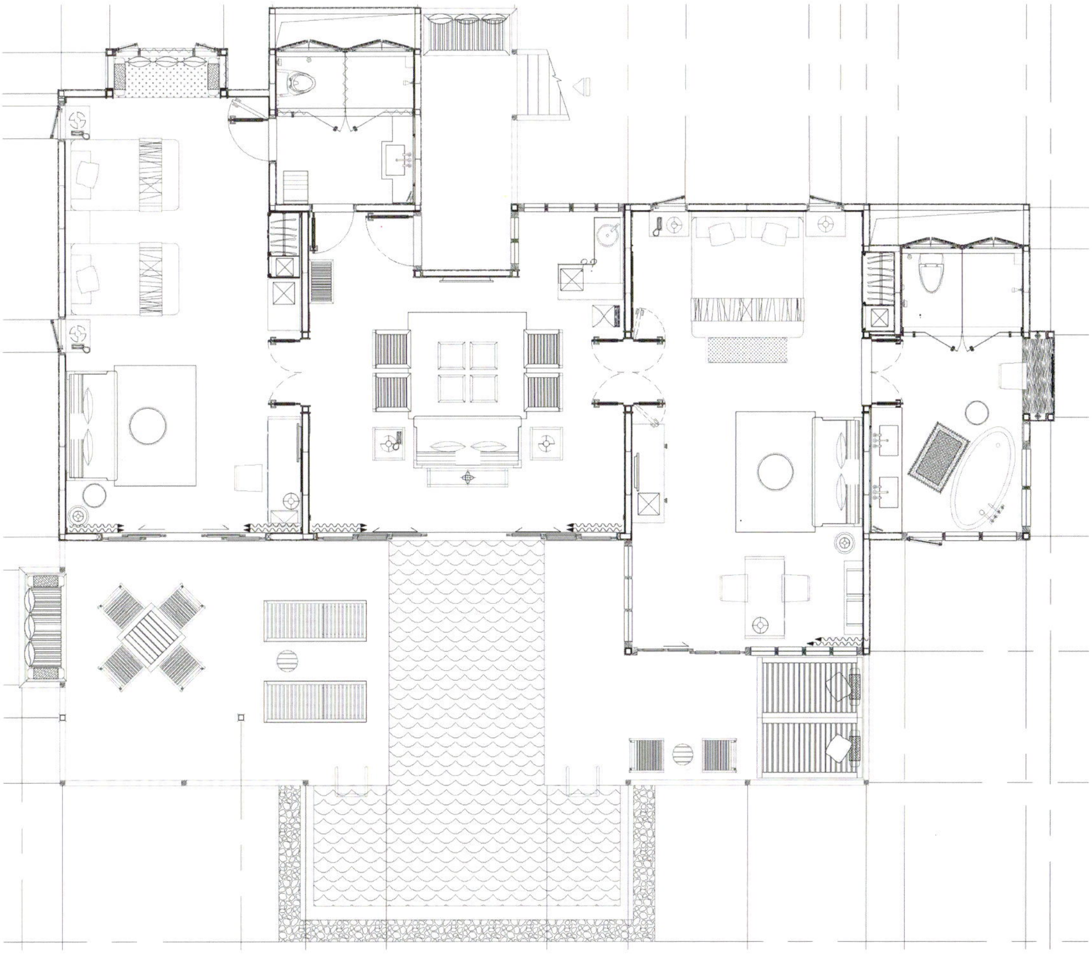

01

地点	面 积	设计师	设计公司	主要材料
/ 南京	/ 3000m²	/ 李宽喜	/ 品奕装饰设计工程有限公司	/ 实木线条混水漆、横截面木饰、玻璃、银镜、乳胶漆、环保亚麻地板、大理石、定制现代工艺吊灯、壁炉、简欧家具、装饰画、布艺

Nanjing Han Seng Hotel Express

南京汉之森商务酒店

The design theme of Nanjing Han Seng Hotel Express is the concept of environmental protection and low-carbon life, which combines the environmental protection with nature. The project is located by Baijia Lake in Jiangning District (Nanjing), an area for fine-quality life, where such a design philosophy is more necessary. Green color runs through the building from the outside to the inside and on the inside, traditional Jane European style is mixed with modern elements. White is used as the background color and is dotted with dark green and navy blue to make the space calm and relaxing, and yet the floor in the lobby and the warm-toned light source add some warm atmosphere to form an appropriate contrast between cool and warm tones. The hotel combines the concept of environmental protection perfectly with the comfort level of modern hotels. The mosaic patterns, the semi-circle information desk, the Jane European style fireplace, the modern style lights...all these offer the hotel a unique effect of mix-and-match.

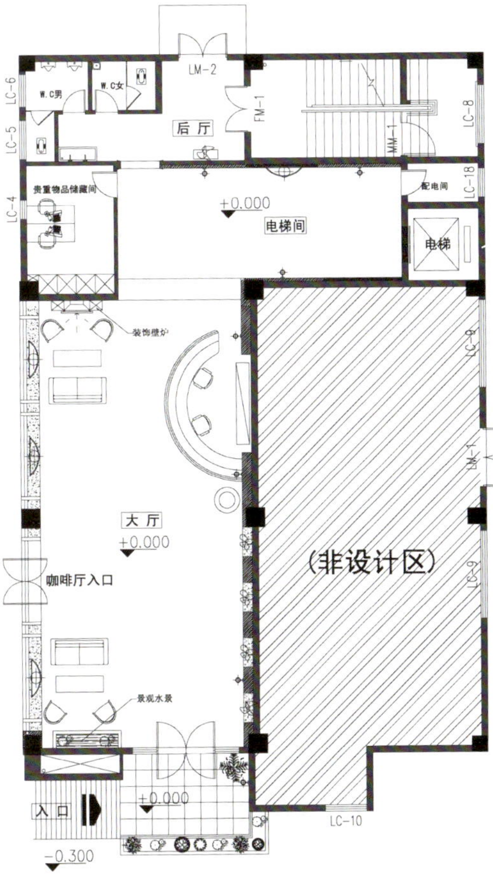

W.C男
W.C女
LM-2
后厅
FM-1
MM-1
LC-6
LC-5
LC-8
LC-4
LC-18
贵重物品储藏间
配电间
+0.000
电梯间
电梯
装饰壁炉
LC-9
LM-1
大厅
+0.000
咖啡厅入口
(非设计区)
LC-9
景观水景
入口
+0.000
LC-10
-0.300

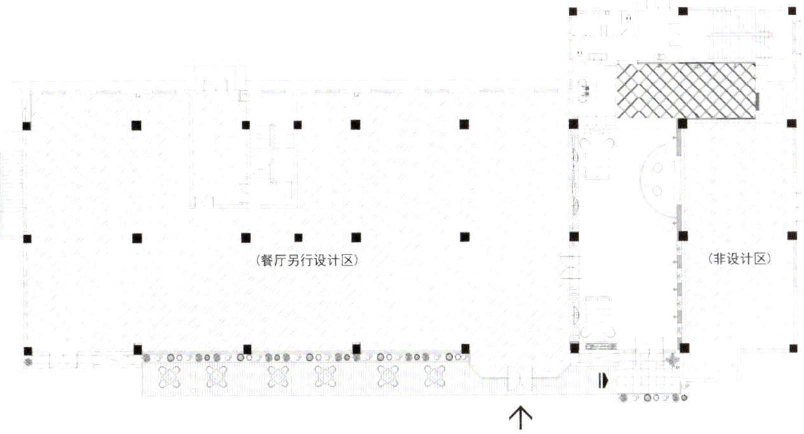

(餐厅另行设计区)
(非设计区)

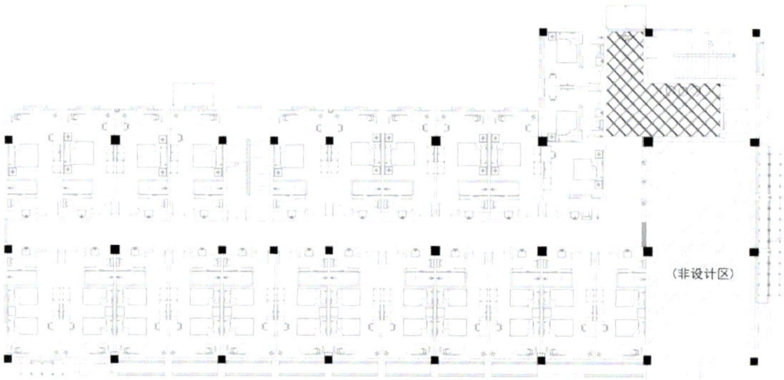

（非设计区）

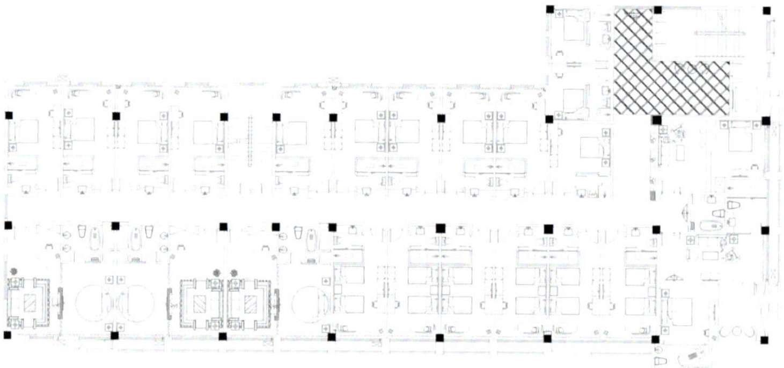

汉之森商务酒店的设计主题是环保的理念和低碳的生活，让绿色环保与自然相结合。项目位于江宁百家湖旁，这里是极佳的高品质生活区，更需要倡导这种设计理念。建筑由外到内贯穿绿色系，室内用传统的简欧风格和现代的元素混搭。白色调作为底色，墨绿色和海军蓝色作点缀，使空间略为冷静、放松。大厅地板和暖色的光源又增添了一些温暖的气氛，冷暖对比恰当。酒店将环保理念与现代酒店的舒适度完美结合。马赛克拼花，半圆形的服务台，简欧式的壁炉，现在风格的灯饰……给酒店带来独特的混搭效果。

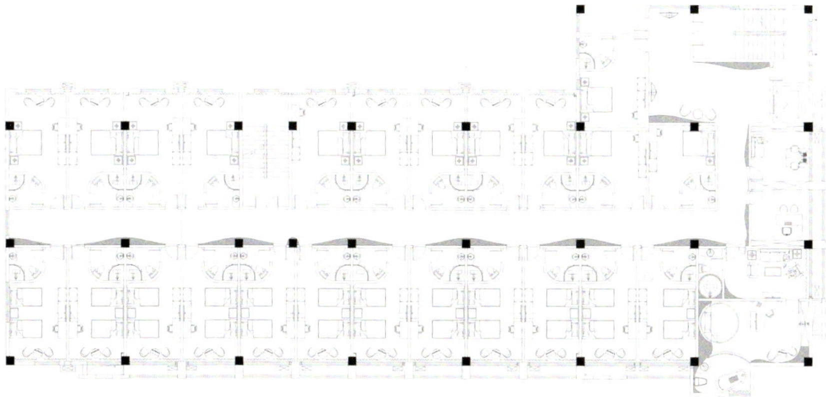

地点 /常州

设计师 / 徐少娴

设计公司 / Gotomaikan International Limited

Howard Johnson Kaina Plaza Changzhou

常州凯纳豪生酒店

The hotel is a three-storey building with lobby, lounge and 24-hour restaurant on the first floor, the Chinese restaurant on the second one and the banquet hall, the third. Based on its function and nature, each floor possesses unique features and brings people various enjoyment of vision and taste.

The flowing lines on the ceiling of the lobby paired with the dazzling light from the light boxes echo with the molding of the elevation wall, creating a magnificent visual effect. The waterscape in the center of the lobby add unique liveliness and elegance to the whole space; the designs and lines at the reception center are relatively simple, and all the design seem to be narrating the journey of the guests.

Specialty restaurants of different flavors show a distinctive character of the hotel. With different functions, the restaurants maintain a unified style as a whole and at the same time, carry special features through different design elements, bringing people visual enjoyment of a higher level.

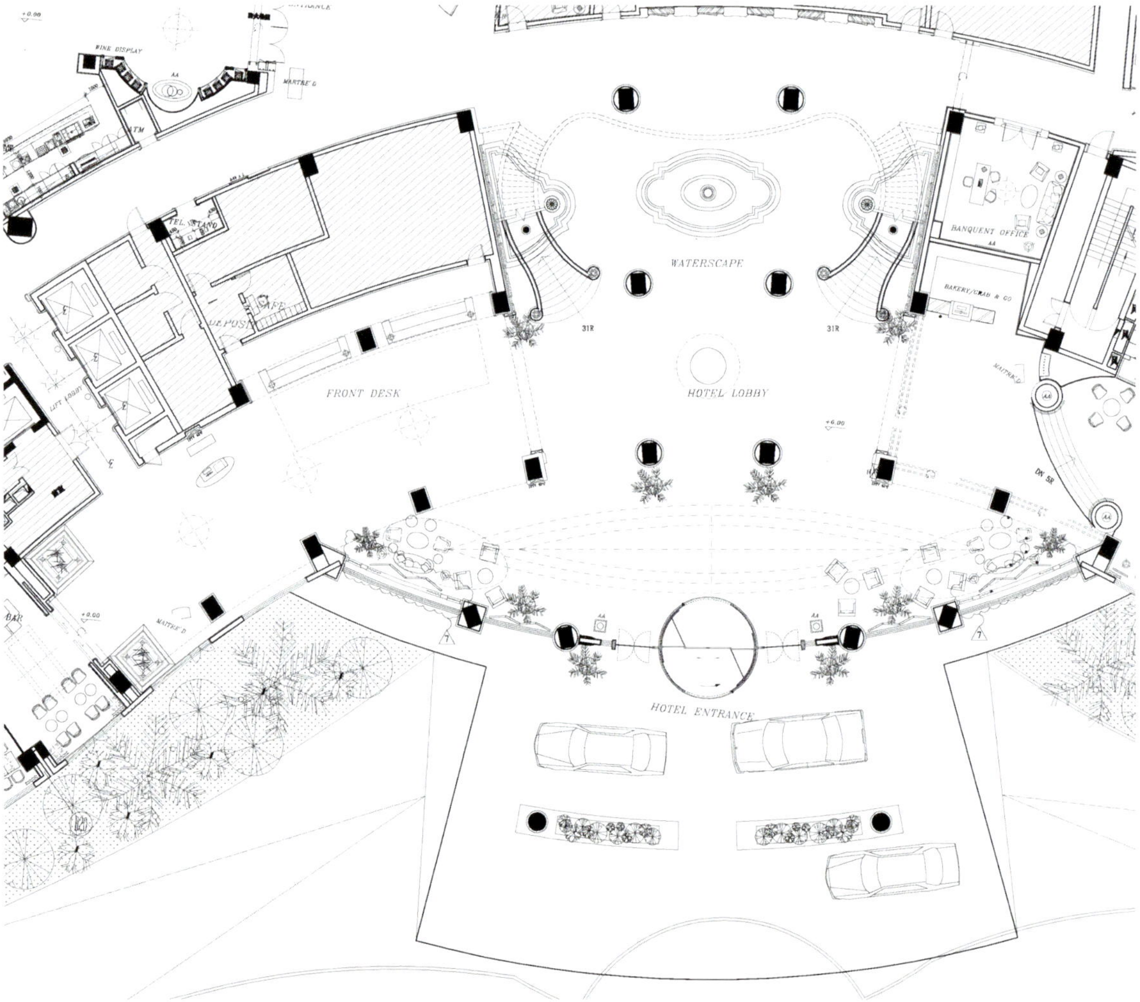

WINE DISPLAY
WATERSCAPE
BANQUET OFFICE
FRONT DESK
HOTEL LOBBY
HOTEL ENTRANCE

豪生閣
LOTUS GARDEN

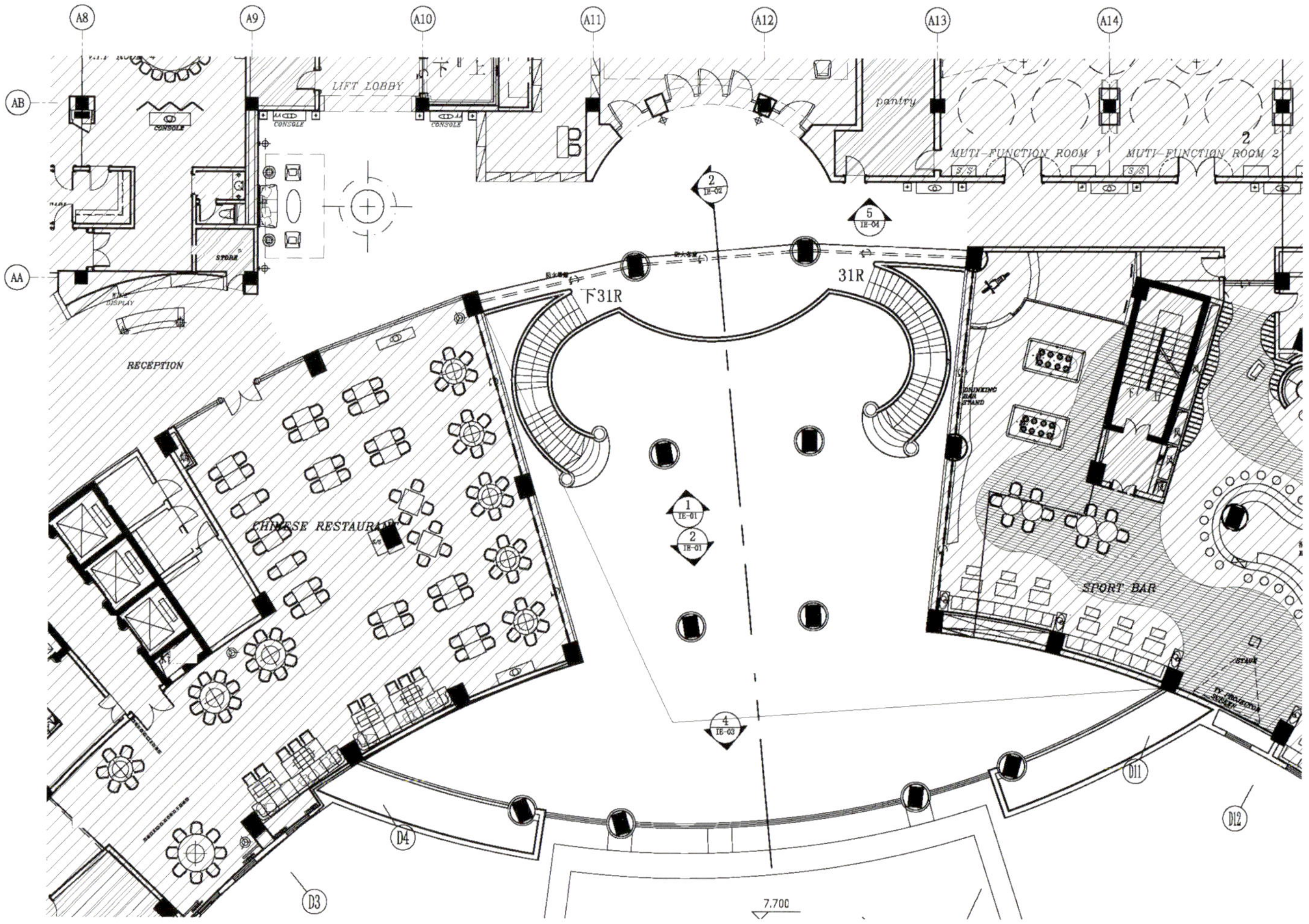
A8
A9
A10
A11
A12
A13
A14
AB
AA
LIFT LOBBY
CONSOLE
pantry
MUTI-FUNCTION ROOM 1
MUTI-FUNCTION ROOM 2
STORE
RECEPTION
F31R
31R
CHINESE RESTAURANT
SPORT BAR
D3
D4
D11
D12
7.700

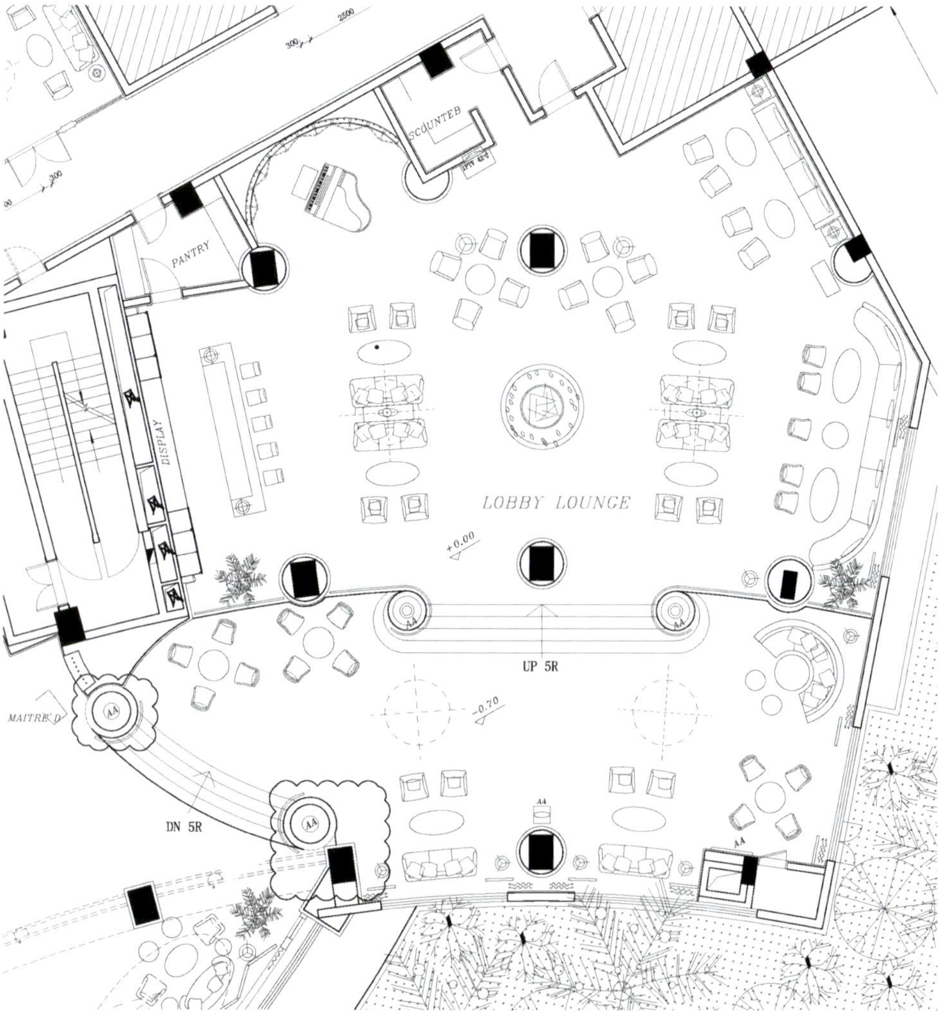
PANTRY
DISPLAY
LOBBY LOUNGE
+0.00
UP 5R
-0.70
MAITRE'D
DN 5R

安全出口

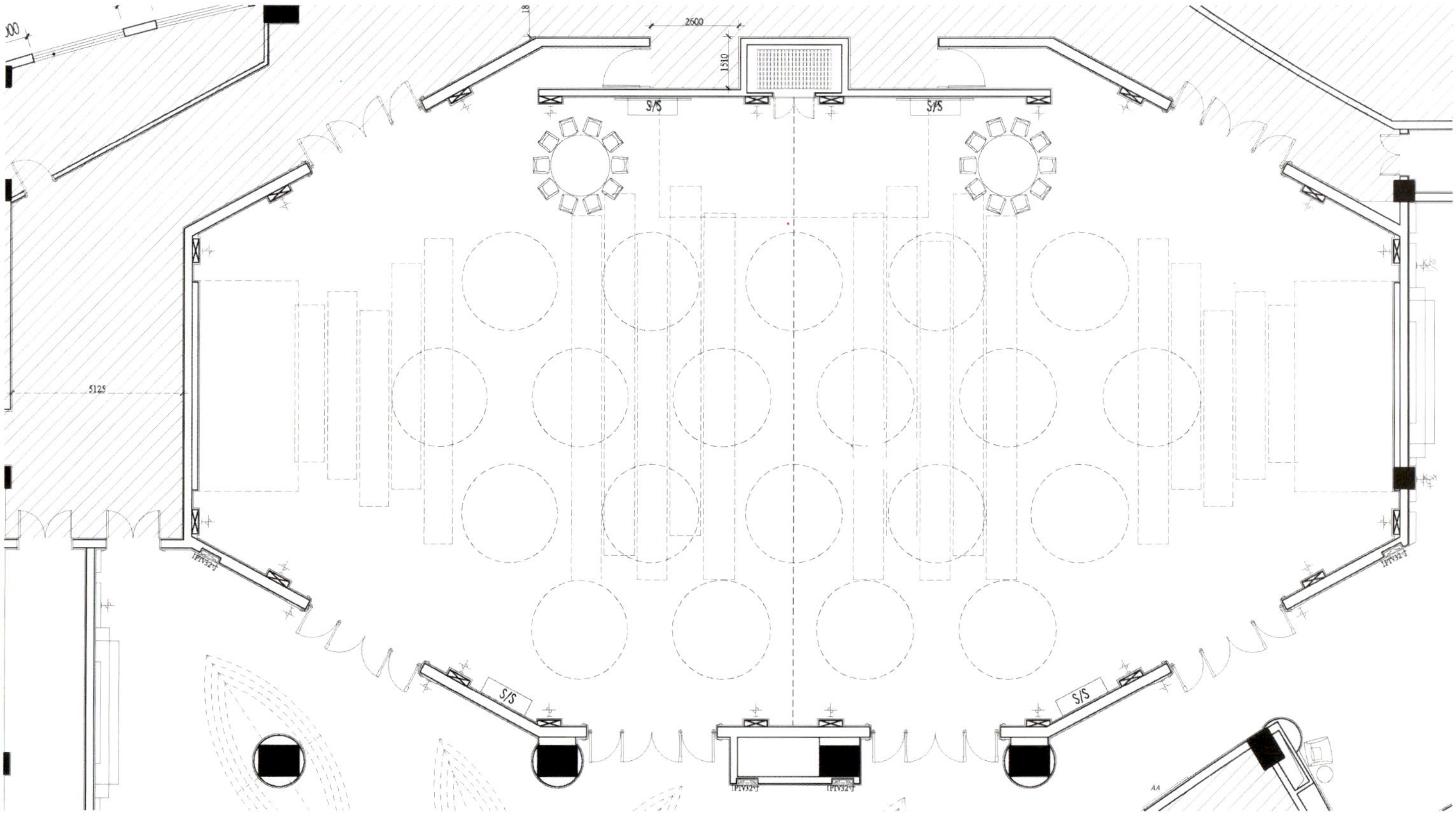

2600
1510
5125
S/S
S/S
S/S
S/S

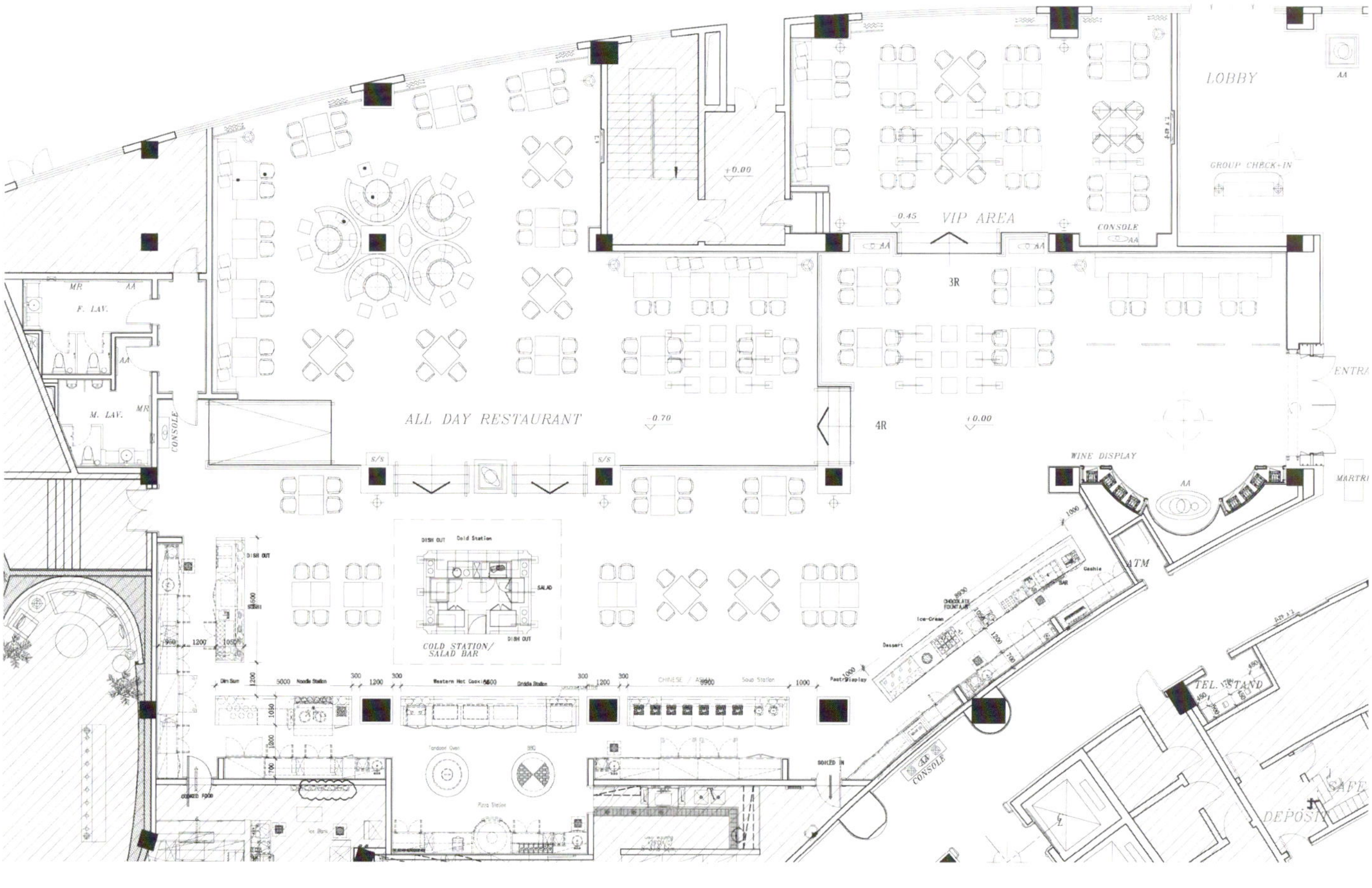
ALL DAY RESTAURANT
VIP AREA
LOBBY
GROUP CHECK-IN
CONSOLE
WINE DISPLAY
COLD STATION/
SALAD BAR
M. LAV.
F. LAV.
ATM
TEL. STAND
3R
4R

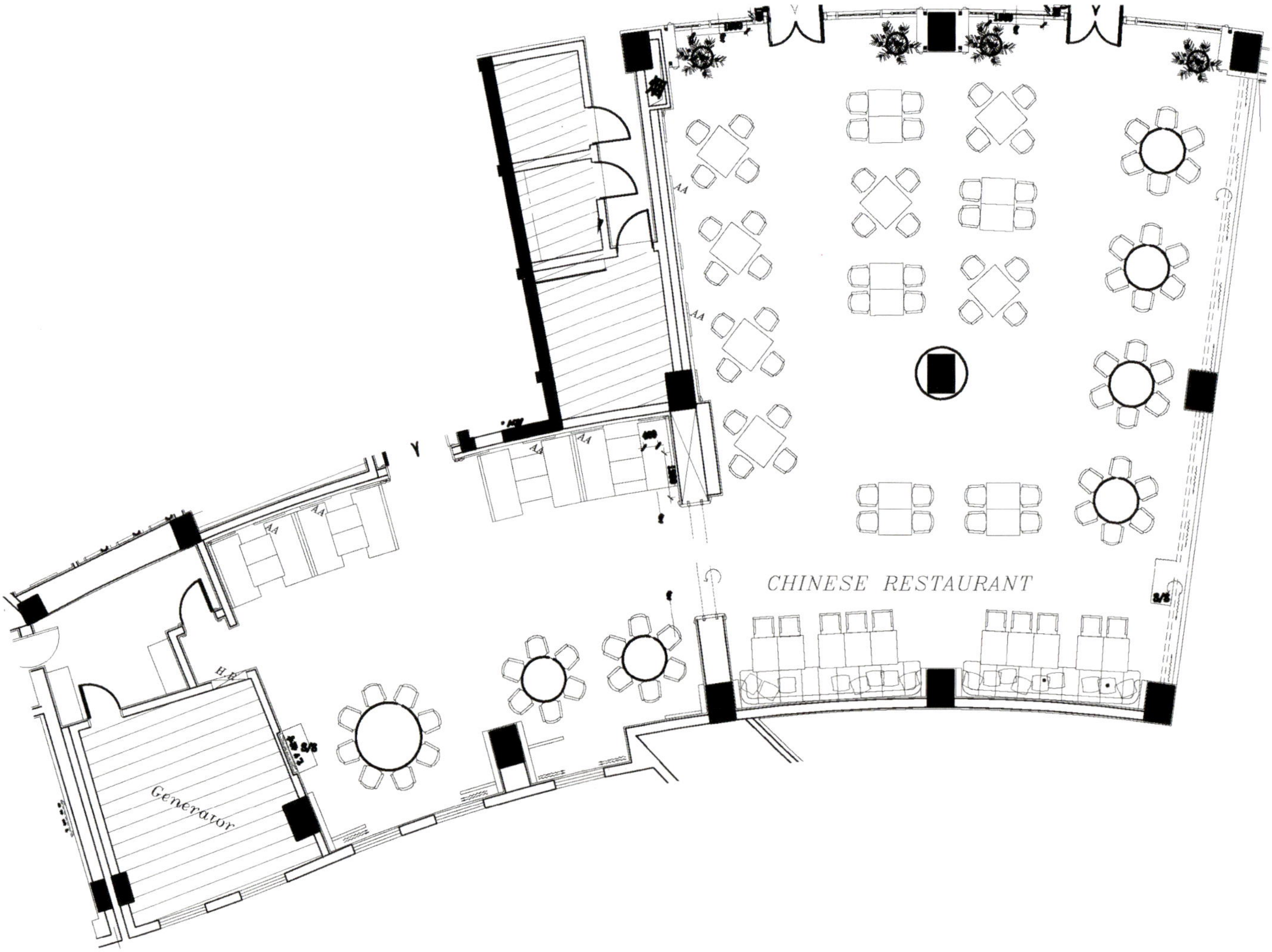

CHINESE RESTAURANT
Generator

酒店共有三层，一层是大堂、酒廊、全日餐厅；二层是中餐厅；三层是宴会厅。每一层都因其不同的功能性质而各具特色，带给人视觉、味觉等多方位的感官享受。

大堂天花线条流畅，配以发光箱夺目耀眼的光芒，与立面的墙体造型相呼应，营造出富丽堂皇的视觉效果。位于大堂中央的水景设置为整体空间增加了生动而优雅的独特氛围；接待处相对简约的设计及线条似乎都在讲述着客人们的行程。

酒店不同风味的特色餐厅使整个酒店彰显了其独特性。不同功能的餐厅在整体风格上保持统一，同时又通过不同的设计元素使每个空间都别有特色，带给人更高层次的视觉享受。

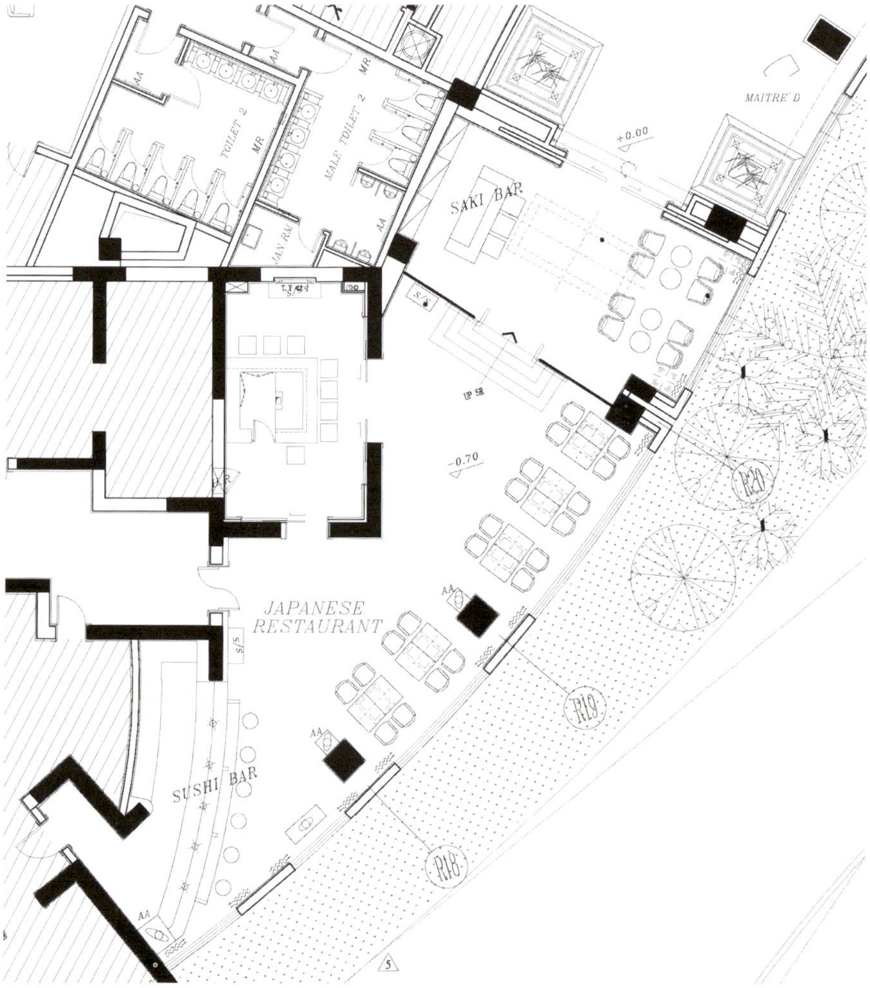
TOILET 2
MR
MALE TOILET 2
MR
JAN/RM
SAKI BAR
MAITRE' D
+0.00
UP 58
-0.70
JAPANESE
RESTAURANT
SUSHI BAR
R18
R19
R20

地点	面 积	景观设计	深化设计	设计公司	主要材料
/ 苏州	/ 5500m²	/ 王海鹏	/ 王钰博	/ 沈阳自然景观艺术有限公司	/ 绿可木、天然木材、雕塑、花岗岩、大理石

Suzhou Wuyue Ducheng Ecological Grand Hotel

苏州吴越都城生态大酒店

Wuyue Ducheng Ecological Grand Hotel is located in the area of Qionglong Mountain, a renowned resort area in Suzhou city and also the birthplace of *The Art of War* by Sun Wu. The site area of the hotel is more than 100 mu and with its back facing the beautiful Taihu and Qionglong Mountain which has natural sceneries. In here, there are tall, ancient trees, eye-dazzling rare flowers, bridges and running water, relaxation and catering areas. Abundant green colors can be seen everywhere.
The design theme of the project is to regard the water as the pulse, green colors as the soul, and history as the root to carry forward Wuyue culture and exemplify the ecological garden landscape culture. The design is divided into an anteroom, a castle of Liangzhu culture, Sunwu Garden, King Wu's Palace and tropical rain forests area. In the design of the entrance, the historical background of Kingdom Wu and Yue Hegemony is introduced, which is demonstrated by classic icons of Chinese culture such as azure dragons, white tigers, rosefinches, Xuanwus and banners. The whole design turns the Liangzhu culture with a history of thousands of years into simple and unsophisticated cultural heritages, which touches the visitors deeply.

吴玉阁

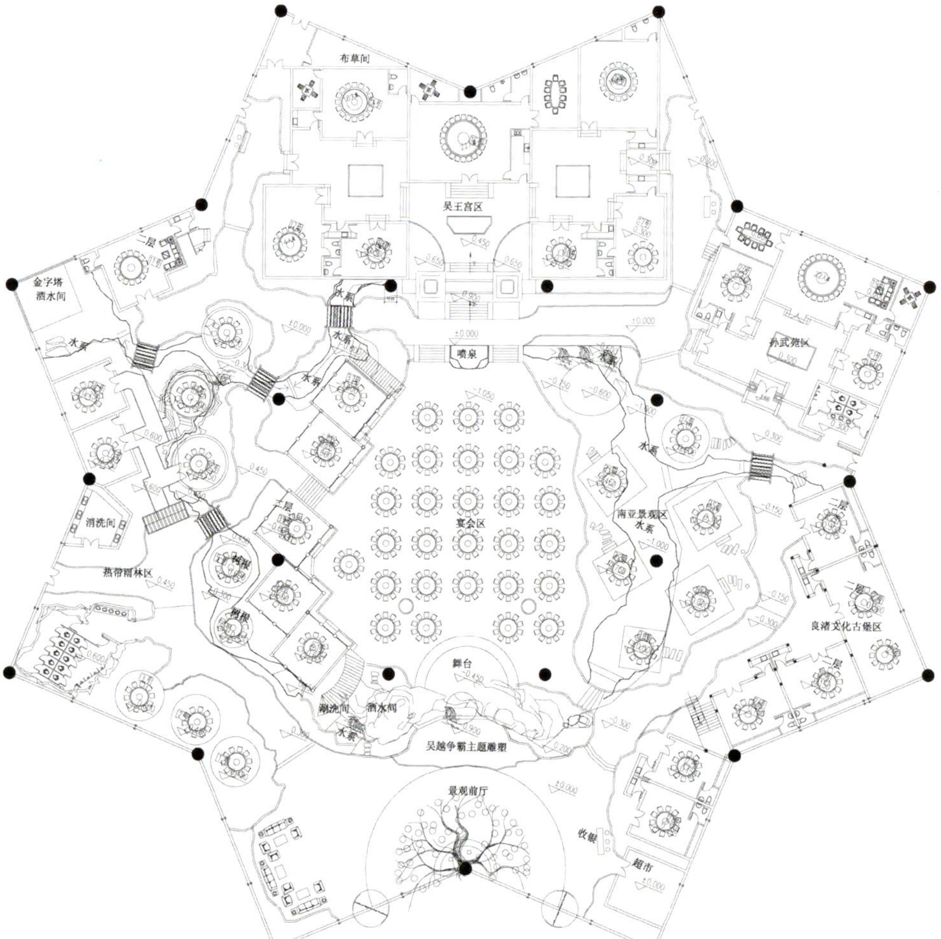

吴越都城生态大酒店位于苏州市著名风景旅游区——《孙子兵法》的诞生地穹窿山，酒店占地百余亩，背倚美丽的太湖之滨、自然风景秀美的穹窿山。这里古树参天，奇花夺目，小桥流水，休闲餐饮尽在其中，处处绿意盎然。

本案的设计主题以水为脉、以绿为魂、以史为根，弘扬吴越文化，体现生态园林景观文化。在设计上划分为前厅、良渚文化古堡、孙武苑、吴王宫、热带雨林区。在入口的设计上引入吴越争霸的历史背景，以中国文化经典的代表——青龙、白虎、朱雀、玄武、战旗等文化元素简化而成。整个设计将积聚千年的良渚文化遗存化作古朴沧桑的文化遗迹，使观者心灵震撼。

地点	面积	设计师	设计公司	主要材料
/ 佛山	/ 1568m²	/ 杨铭斌、李嘉辉、何晓平	/ C.DD（尺道）设计师事务所	/ 乳胶漆、玻璃、木饰面、不锈钢

Homyip Hotel

红叶坊精品酒店

The theme of this project is "makeunder", a makeunder of nature. Each image reflects the beauty of nature: woods–the projection wall in the waiting area of the lobby, falling leaves–the glass curtain wall in the business area of the lobby, dripping water–the background curtain wall in the recreation area of the guest rooms floor, veins of the earth – the bedside background in the theme room, grassland–the parquet patterns in the corridor carpet of the guest floor. What the guests get is a feeling of being natural and relaxed, and here they can stay close to nature after a busy working day; a deep breath or a visual sensation of a makeunder can always give you the enjoyment of sleeping in the arms of nature.

Besides integrating the space design with concepts, the designers emphasize on the choices and matching of the materials so that both colors and texture effects are rich and attractive. Everything that can be seen generates cordial feelings and supplies the metropolis guests with intimate atmosphere of nature.

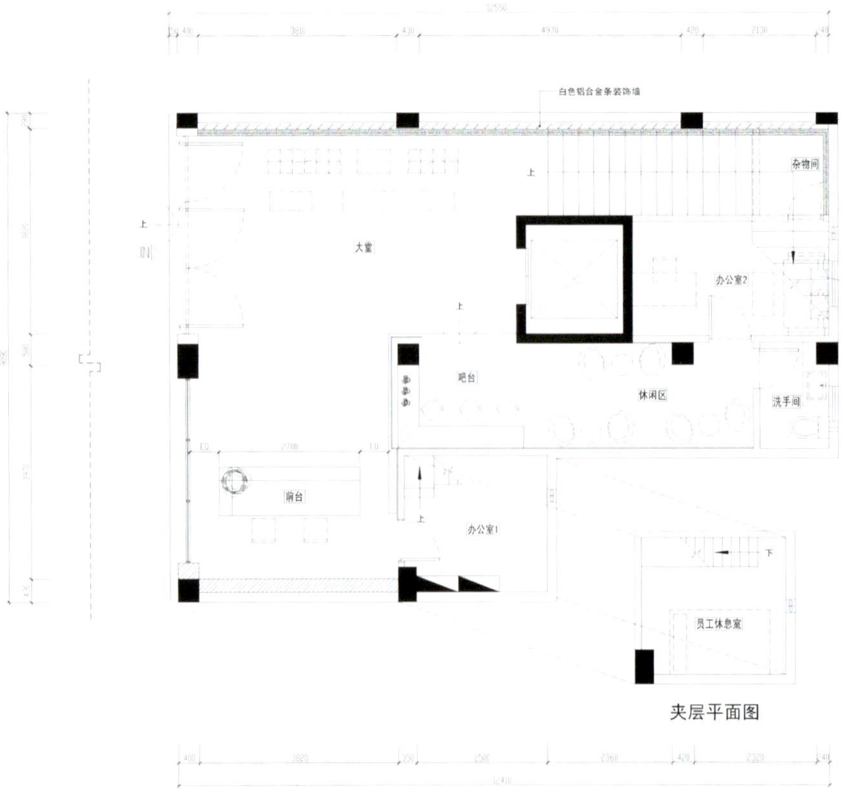

夹层平面图

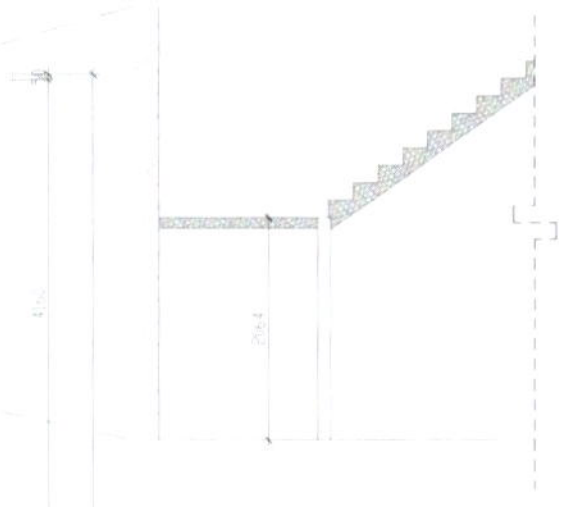

杂物室立面图

本案主题为“素颜”，大自然的素颜。每一处影像都反映着天然之美：树林——大堂等候区投影幕墙，落叶——大堂商务区玻璃幕墙，滴水——客房层休闲区背景幕墙，地脉——主题房间床头背景，草地——客房层通道地毯拼花图案。回馈给客人的是一种天然、放松的感受，让客人在忙完一天工作后，在这里与大自然亲密接触；一个深呼吸，一份素颜的视觉感观，都可以让您享受地熟睡在大自然的怀抱中。

除了酒店空间的规划融入概念外，设计师在物料选取时，注重相互搭配，让材质的色彩和肌理效果都非常丰富且极富美感。视线所及之处均有一种亲切感，让都市客人们感受到亲切的自然气息。

地点	面积	设计师	设计公司
/ 北京	/ $25000m^2$	/ 冯厚华	/ J2-DESIGN顾问设计有限公司

Beijing Xidan Grand Mercure Hotel

北京西单美爵酒店

Accor, a French Hotel Management Group, upholds its Grand Mercure's brand positioning as providing luxury hotel accommodation with local characteristics. The Grand Mercure Hotels, which are renowned for its unique styles and characteristics, harmonize with the local features, and their fashionable and exquisite design, decoration, catering and service are all reflecting the most unique style of the local place. This project is different from the traditional Chinese style. Not only does it adopt Chinese red, long red tables and round-backed armchairs, but also the international appearance of it wraps the feelings and connotation of Beijing, which are presented in a way of spatial installation to the visitors all around the world.

The original space of the lobby is restricted and conservative, so the designer moves the reception desk to the left to make it closely connect with the luggage room and the room for keeping valuables. In this way, the spacious, generous, and circulated lobby space is opened up which guides people's first view to the innermost lobby bar where fresh flowers, nice wines and arts are used to welcome visitors.

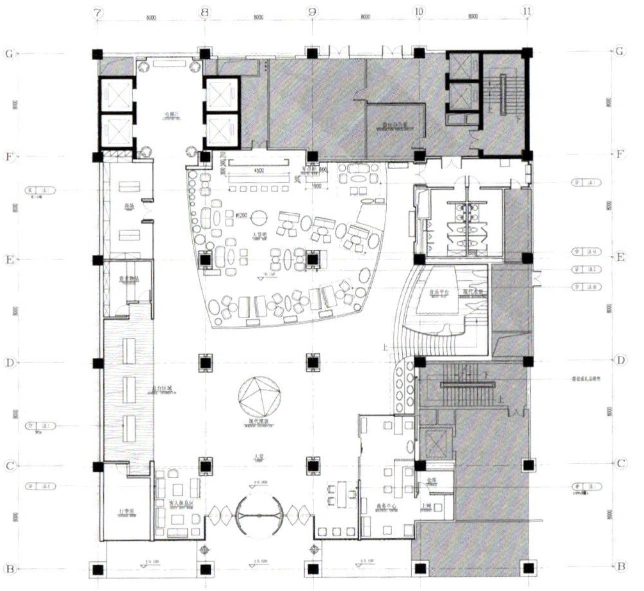

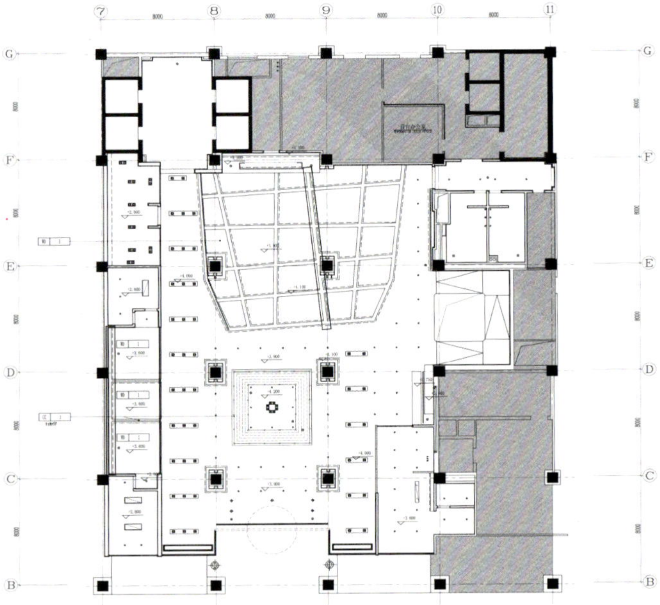

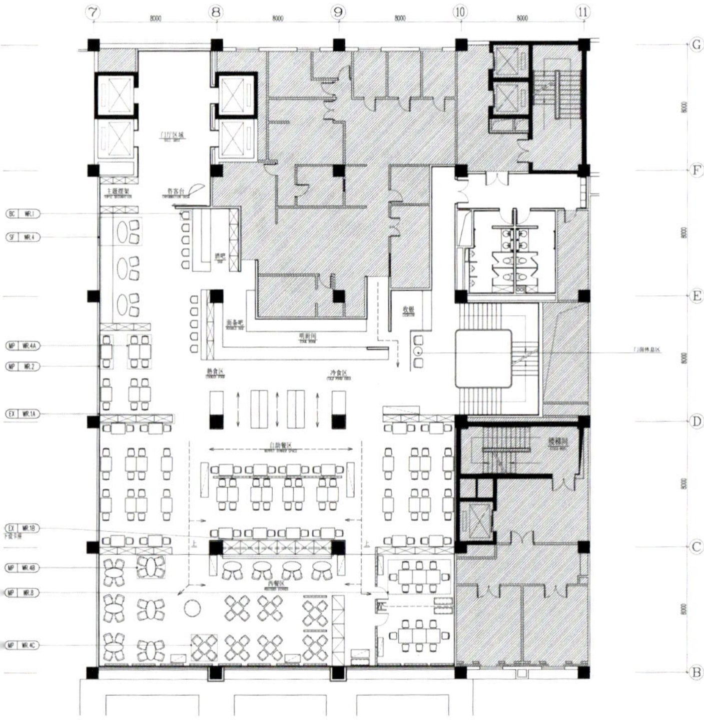

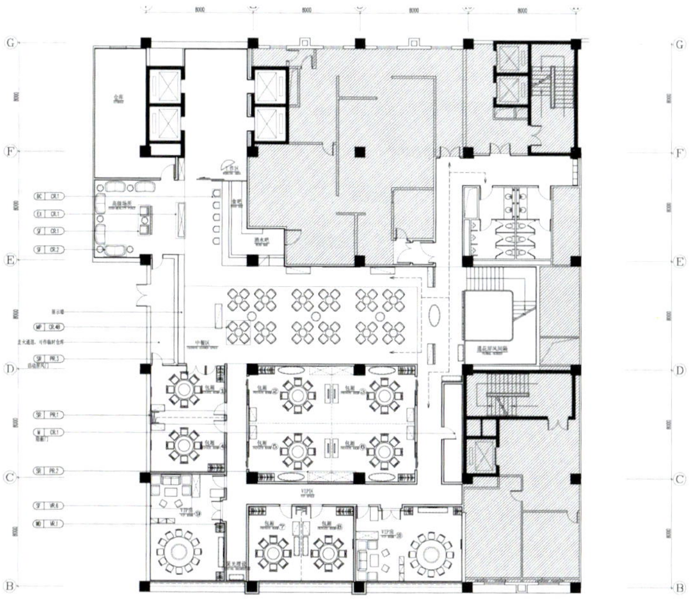

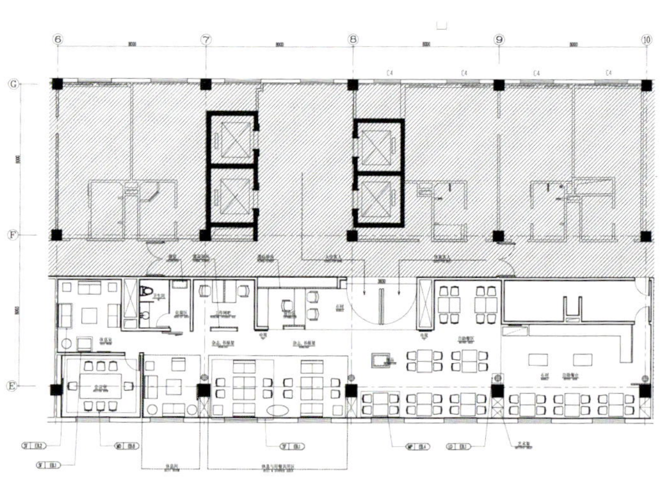

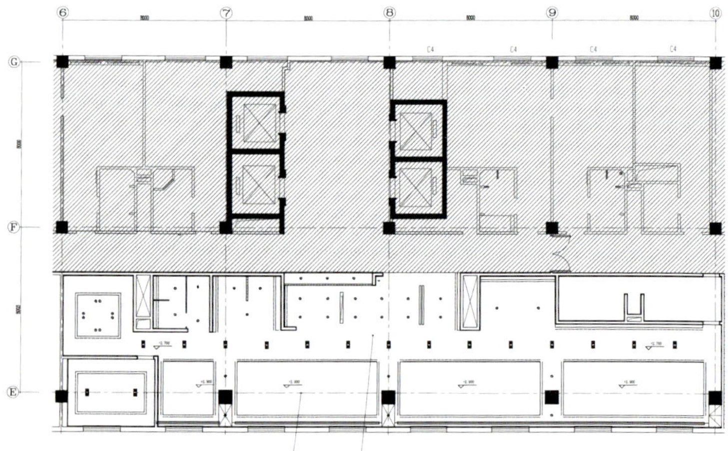

法国雅高酒店管理集团将其旗下的美爵品牌定位为：提供带有地方特色的豪华住宿的酒店。以独特的风格及个性而闻名的美爵酒店都与当地特色相融合，从设计、装潢、餐饮到服务，全都反映出当地最独特的风格，且时尚、精致。本案有别于一贯的中式风格，不仅采用中国红与条案、圈椅，还用国际化的外表包裹北京的情感与内涵，以空间装置的形式呈现给世界的宾客们。

大堂的原空间拘束且保守，设计师将前台移至左侧，与行李间和贵重物品室紧密相连，开拓出宽松、大方、流通的大堂空间，将第一视觉引至最里面的大堂吧，用鲜花、美酒和艺术来迎接宾客。

地点	面积	设计师	设计公司	摄影师
/ 福州	/ 8505m²	/ 于丹鸿	/ 重庆朗图室内设计有限公司	/ 施凯

MUSE City Hotel

沐思城市酒店

MUSE City Hotel is popular among customers for its unconventional mode of services. The special services it offers, such as sufficient time for check-in, exclusive floors for ladies, full range of business facilities and the sharing of original source of the hot spring, provide a feeling of home for customers in this metropolitan city. Adhering to such management system and business mode, the indoor decoration shows warmth and intimacy instead of gorgeous and luxurious style. With total sincerity, the hotel provides you a home away from home, sweet, lovely and warm.

In here, modern minimalist style matches with wooden veneer and flooring, and pleasant colors and lighting add more softness. Simple and magnanimous marble and wood-grain-like flooring are used in public areas, lobby and showering area, showing not only conciseness and grandness, but fashion as well. People feel relaxed here without any loss of style.

MUSE City Hote

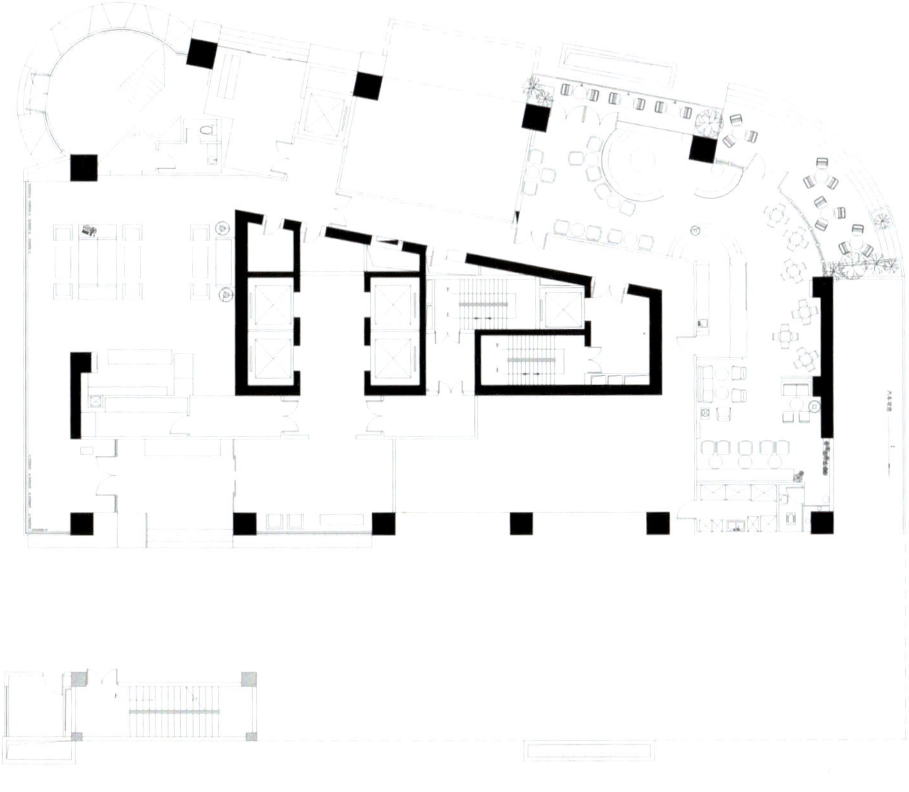

沐思城市酒店因其不落俗套的服务模式而受到顾客的欢迎。充裕的入住时间、专属的女士楼层、全方位的商务配套设施、原脉温泉共享等与众不同的服务方式，为顾客倾心打造城市里的家的感觉。秉承这样的管理制度和经营模式，店内的设计也以温馨、亲近为主，摒弃华丽、奢侈的酒店风格，用心营造一种类似于家的氛围：亲切、优雅而温馨。

在这里，现代简约的风格搭配木饰面和木地板，宜人的色彩和照明更为其增添了几分柔和。公共区域、大厅和卫浴区选择简约、大气的大理石地板或仿木纹地板，简洁、大气之余，也平添了几分时尚气质，让人感觉轻松而不失格调。

地点	面积	设计师	参与设计	设计公司	主要材料
/ 河北	/ 4500m²	/ 霍庆涛、丁洁	/ 汤善盛、赵洪程	/ 北京大石代设计咨询有限公司	/ 白木纹石材、白影木饰面、钨钢、素色壁纸

Huanghua Boyang Hall

黄骅泊阳会馆

Situated around Bohai, Huanghua is a newly-emerging port city. Here diet as well as culture is closely related to the sea. The whole building consists of five floors. The first two floors are high-end hot pot restaurant, with the other three being guest rooms. Boyang means parking in the east where the sun rises, with the implication of flourishing business, therefore here comes the theme of "clear water and beautiful plants".

Here, water is solidified into a variety of patterns in the space. The crystal water droplets pouring down from the top are instantly captured into images, forming a transparent crystal wall—a light pool with flashing lights and water droplets.

The use of a large amount of flowing water landscape adds liveliness to the space. That down-pouring waterfall wall, overlapping winding water, and out-booming fountain...while decorating the space, all provide good regulation to the dry indoor environment in the North. The use of light-colored wood veneer and wood grain stone adds brightness and warmth to the whole environment. The soft natural wood grains echo with the crystal water droplets, creating the scene of "clear water and beautiful plants".

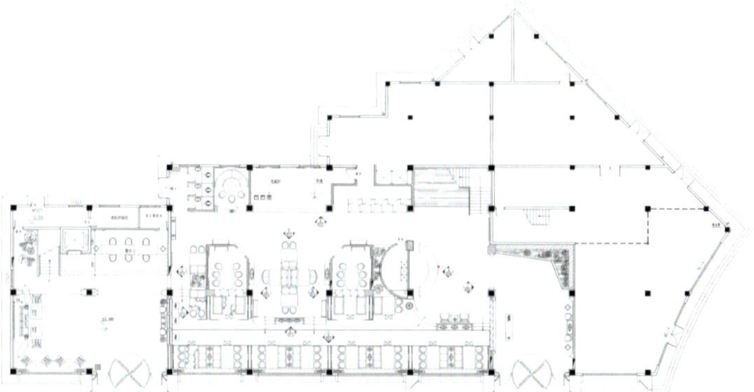

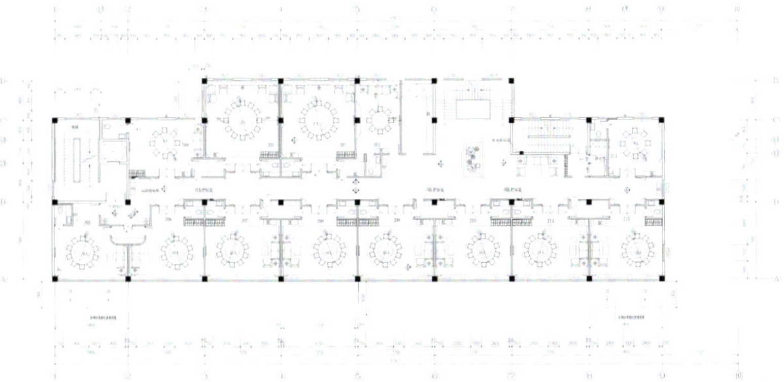

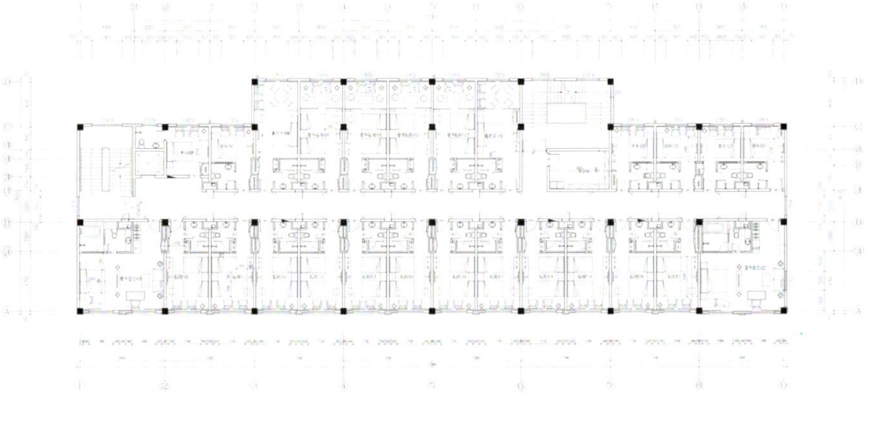

黄骅是新兴环渤海港口城市，无论是饮食结构还是文化都与海有着紧密的关系。整幢建筑共五层，一、二层是高档火锅餐厅，上面三层是客房。“泊阳”取意停泊在东方太阳升起的地方，寓意事业蒸蒸日上，因此才有了“水木清华”这个主题。

在这里，水被空间凝固成多种形态。从顶部倾泻而下的晶莹剔透的水滴，被瞬间捕捉成影像，形成了那晶莹剔透的墙——光影烂漫的水滴灯池。

大量活水景观的运用为空间增添了许多灵气，那倾泻而下的瀑布墙，蜿蜒曲折的叠水，蓬勃而出的喷泉……在为空间提供装饰的同时，也为北方干燥的室内环境起到了很好的调节作用。浅色木饰面和木纹石材的运用，赋予整体环境明亮、温馨的色调。柔美的天然木纹与晶莹的水滴交相辉映，形成了一幅“水木清华”的画面。

深圳港

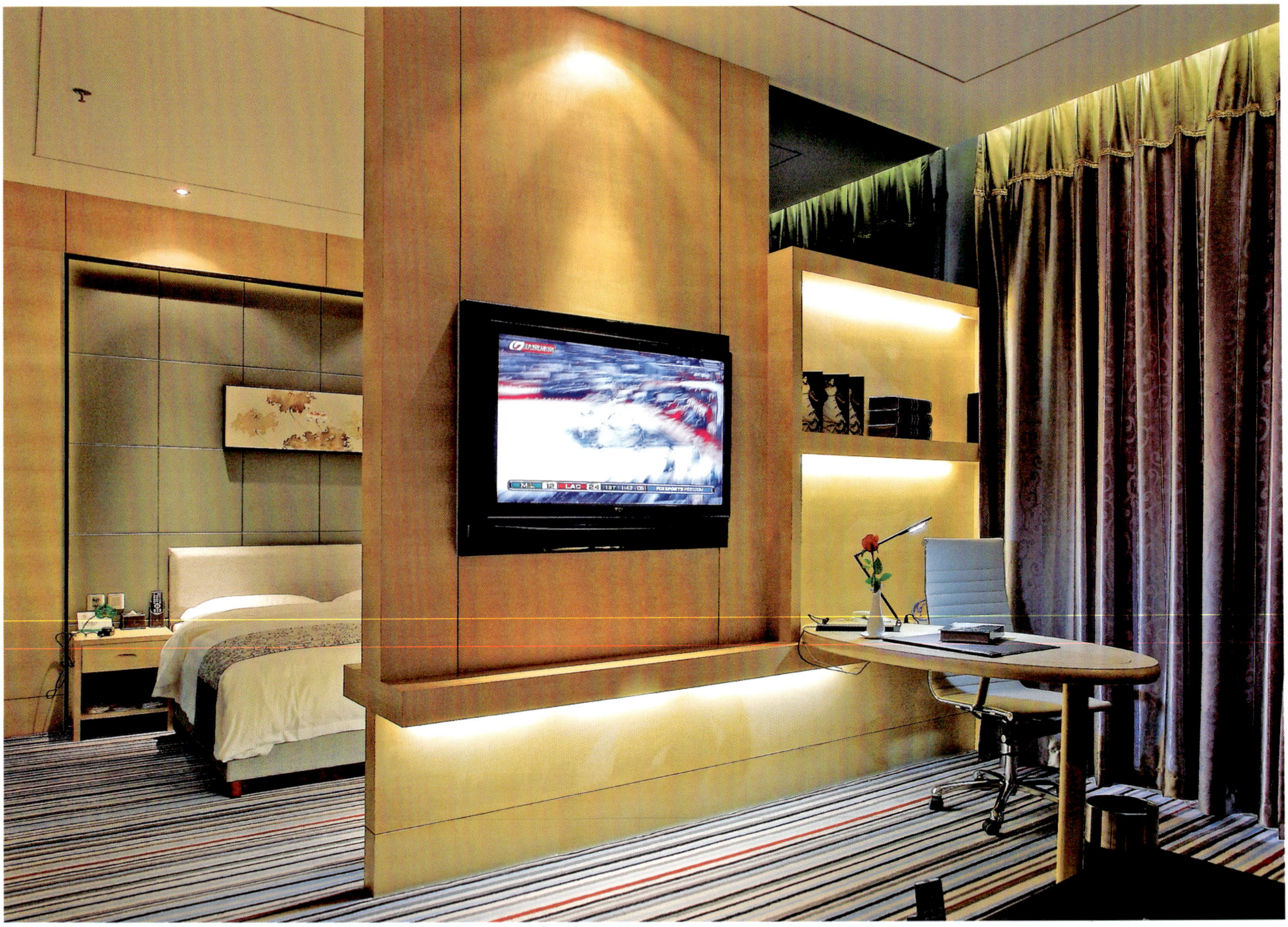

地点	面积	设计师	设计公司	主要材料
/ 河南	/ 15000m²	/ 王军	/ 河南励时装饰设计工程有限公司	/ 云石灯片、新西兰米黄石材、新世纪米黄石材、花梨木饰面、墙纸、皮革硬包

Kaiyuan Zhongzhou International Hotel

开元中州国际饭店

Kaiyuan Zhongzhou International Hotel, a detached building, altogether has nine floors. This detached hotel appears more grand and magnificent in the open area. Fountains are built in front of the entrance square and the water mist, lighting and plants complement each other laying a nice atmosphere foundation for the hotel.

The hotel is internationally oriented and being spectacular, luxurious and elegant are essential qualities of an international hotel. The interior decoration blends the international trends and in the meanwhile it exemplifies the local culture and only in this way can it demonstrate the characteristics of the hotel. The designer considers carefully in aspects of selecting materials and color matching. The classic colors, namely, composed dark, luxurious golden and jumping red have appropriate light and dark shades when group together. Coupled with nice-texture materials, the space revealing more sense of quality and taste, offers high-profiled enjoyment and low-key appealing.

开元中州国际饭店

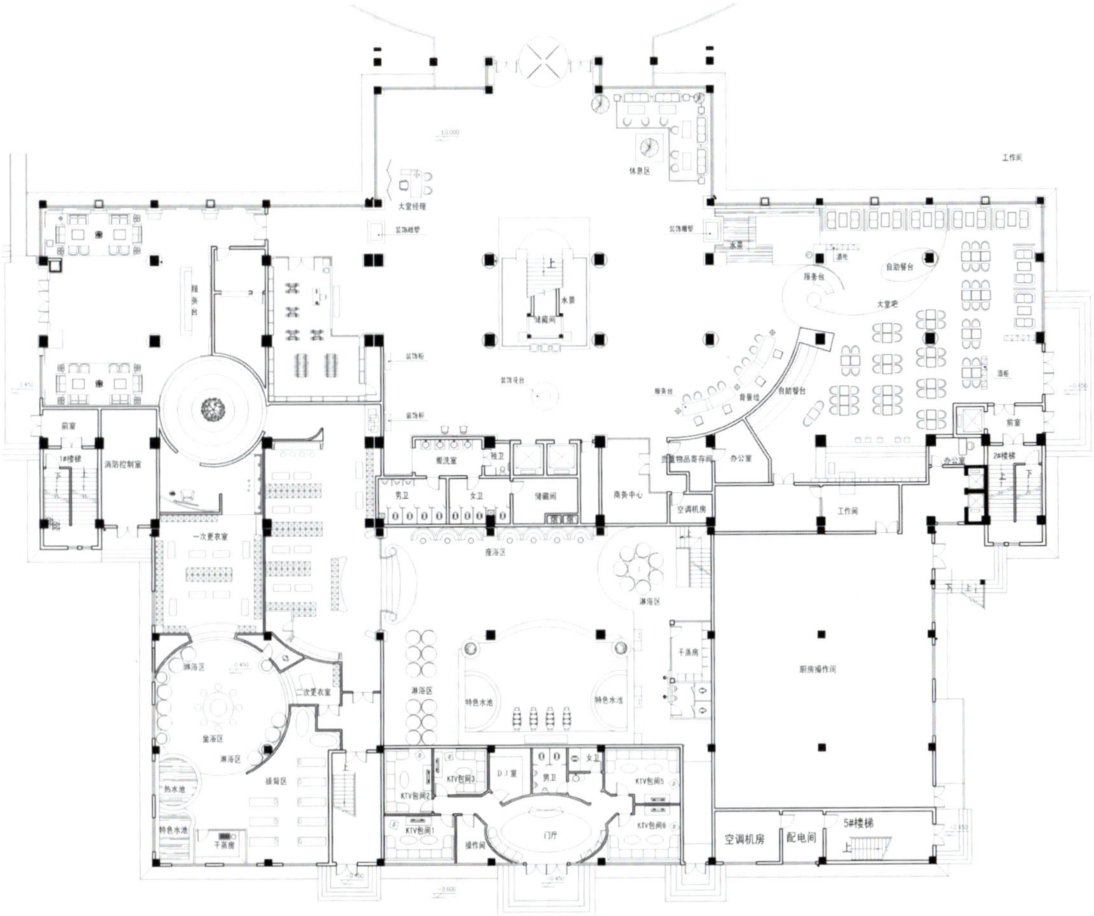

工作间
休息区
大堂经理
装饰墙
水景
储藏间
自助餐台
服务台
大堂吧
酒柜
背景墙
装饰柜
前室
1#楼梯
2#楼梯
消防控制室
盥洗室
残卫
男卫
女卫
商务中心
空调机房
办公室
一次更衣室
二次更衣室
淋浴区
特色水池
热水池
干蒸房
搓背区
厨房操作间
KTV包间1
KTV包间2
KTV包间3
KTV包间5
KTV包间6
D J 室
操作间
门厅
空调机房
配电间
5#楼梯

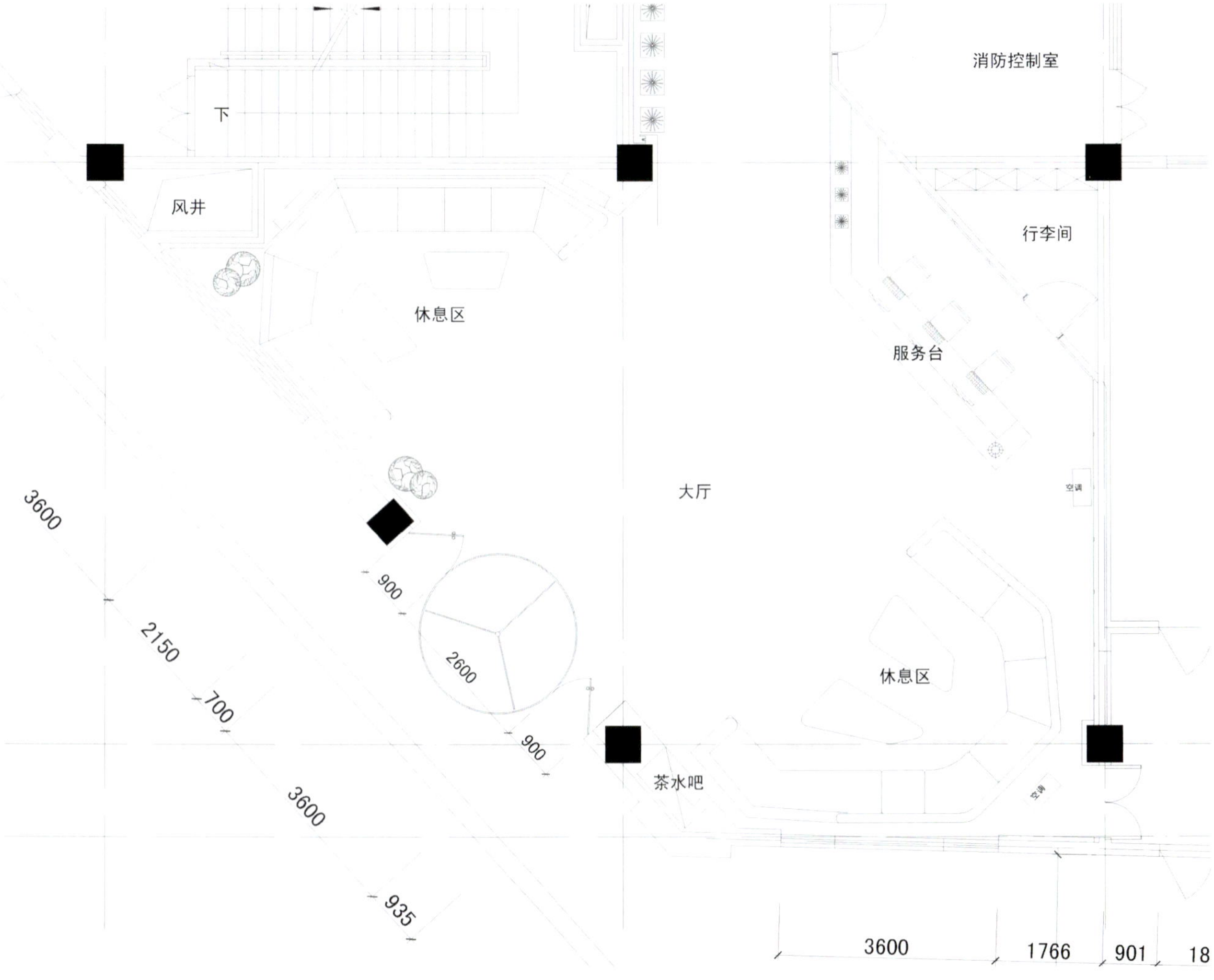
消防控制室
下
风井
行李间
休息区
服务台
大厅
3600
900
2150
2600
700
900
休息区
茶水吧
3600
935
3600
1766
901
18

开元中州国际饭店共九层，属于独立建筑。独栋的建筑在空旷的场地上更显宏大壮观，大门广场前设置了水景喷泉，水雾、灯光、植物，三者相辅相成，为酒店奠定了一个良好的氛围基础。

酒店以国际化定位，恢弘大气、奢华优雅是必备素质。室内设计在融会国际潮流的同时还要体现本地文化，如此方能显出酒店的特色。设计师在选材、色彩搭配方面都经过了深思熟虑，沉着的黑、奢华大气的金、跳跃的红，这些经典色组合在一起，浓淡合宜；加上富有质感的材料，更是显得有分量、有品位，让人感受高调的享受和低调的情趣。

地点	面积	设计师	参与设计师	设计公司	公司网址	主要材料	摄影师
/ 杭州	/ 56000m²	/ 郑仕樑（Ivan Cheng）	/ 崔北亮、陈晓强、唐益超、王晓娜、刘晓峰	/ IVAN C. DESIGN LIMITED	/ www.icdl-hk.com	/ 精选东南亚名贵石材、天然木料、真皮、高级丝绒布匹、环保乳胶漆、光面玻璃、雕花玻璃、玉石	/ 支康

Hangzhou Qiandao Lake Binjiang Hilton Resort Hotel

杭州千岛湖滨江希尔顿度假酒店

Hangzhou Qiandao Lake Binjiang Hilton Resort Hotel is located by the lakeside of Qiandao Lake, which is known for "the most elegant water under the heaven". It stands next to the mountains, faces the lake and is an international hotel with the longest lakeshore line in the area of Qiandao Lake.

With its conceived design focusing on the idea of water, the hotel looks like a magnificent poem. The thick and thin hydrological patterns in different areas sometimes ripple, sometimes float, sometimes drift and sometimes billow softly as they wish. The design is like inclusive water including hues such as golden, red, and coffee and it features individuality and fashionable elements, which palpitates you with excitement. The art ornaments of the new Chinese, Jane European modern and Southeast Asian styles embellish in between with some of them resounding, and some of them crooning, and the cadence is rich and colorful. The materials are exquisite with wide variety, which brings impacts of diversified layout. The design has a style of its own and is not onefold and the diversified touches endow the hotel with diversified wonderlands. Strolling aimlessly in the hotel, it feels like the poetic culture deposits floating in the air. Rambling in the "art gallery" and relishing various arts in the world ,you will feel like you are in a dreamy and fantasy world.

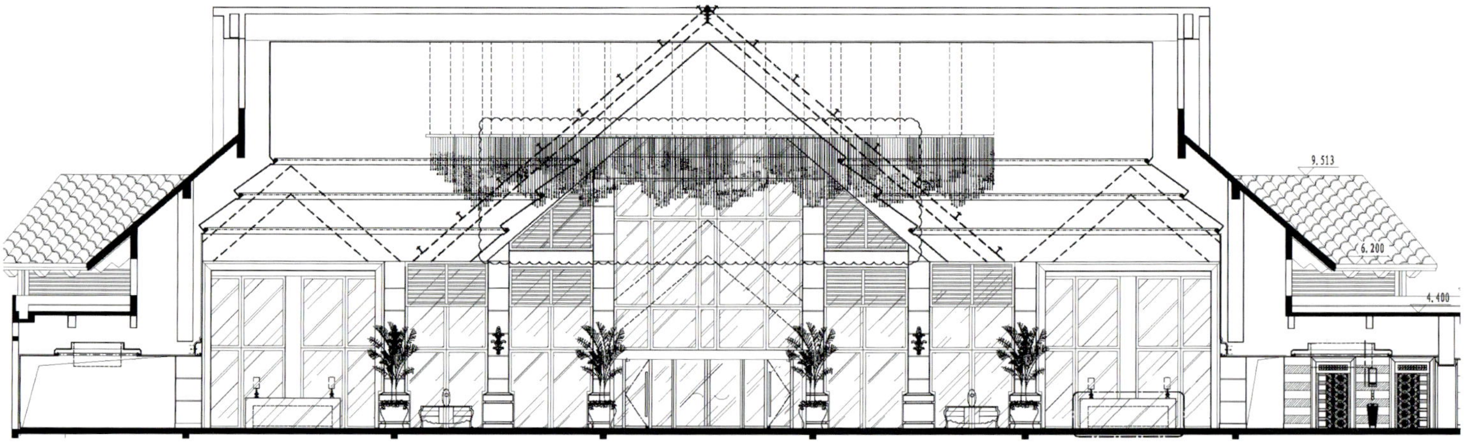
9.513
6.200
4.400

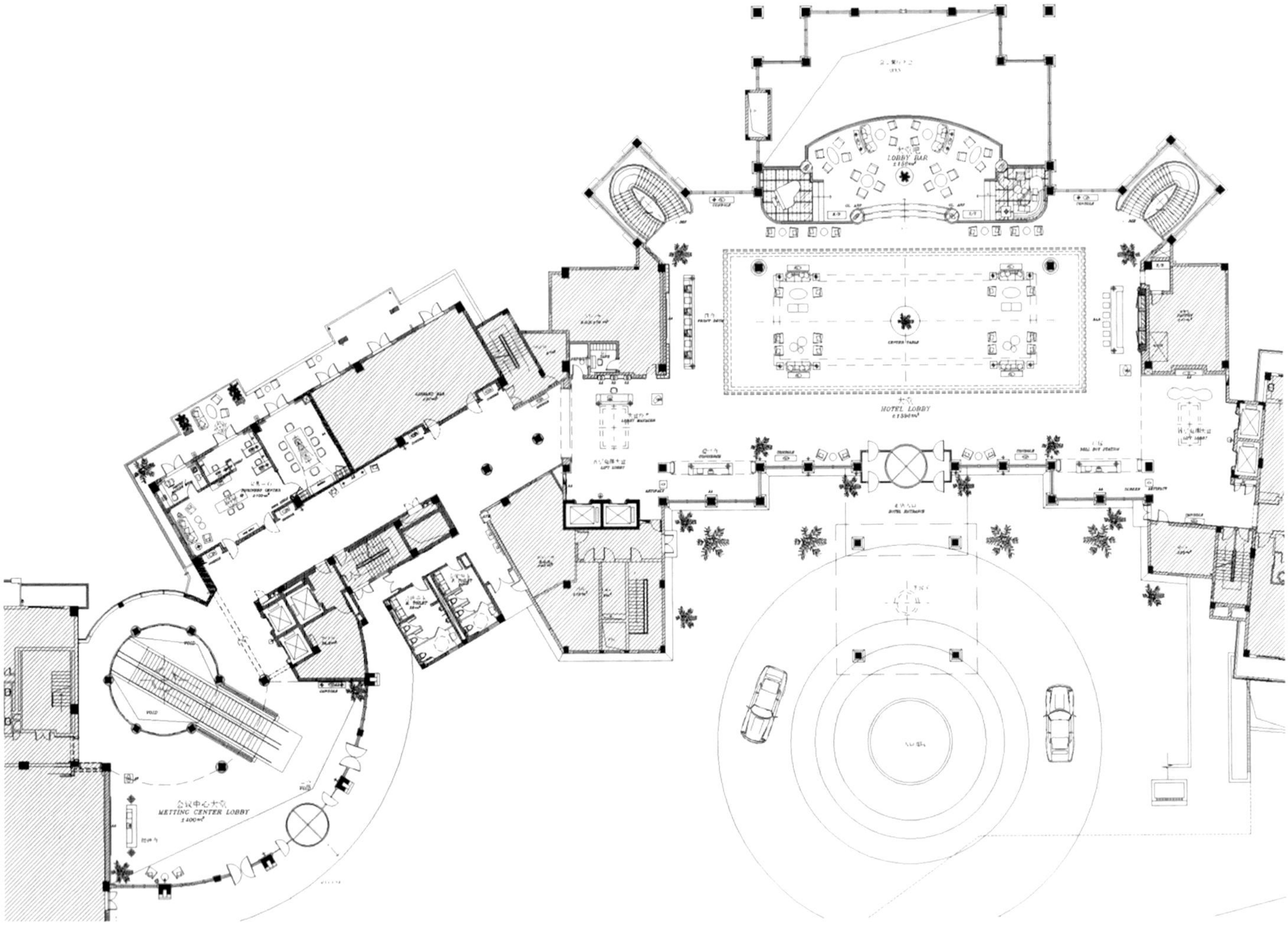
LOBBY BAR
HOTEL LOBBY
会议中心大堂
METTING CENTER LOBBY

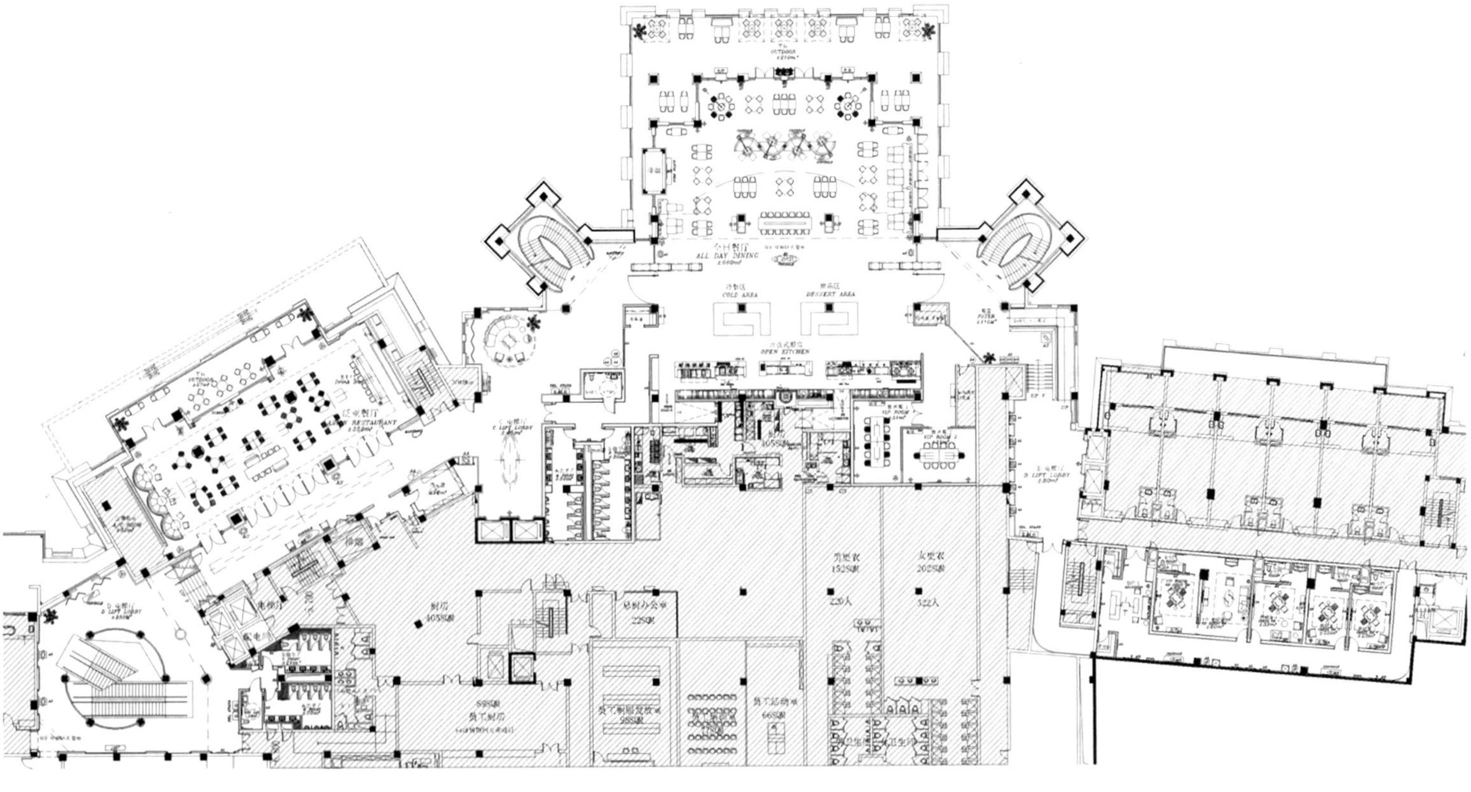
全日餐厅
ALL DAY DINING
冷餐区
COLD AREA
甜品区
DESSERT AREA
OPEN KITCHEN
男更衣
152S㎡
220人
女更衣
202S㎡
322人

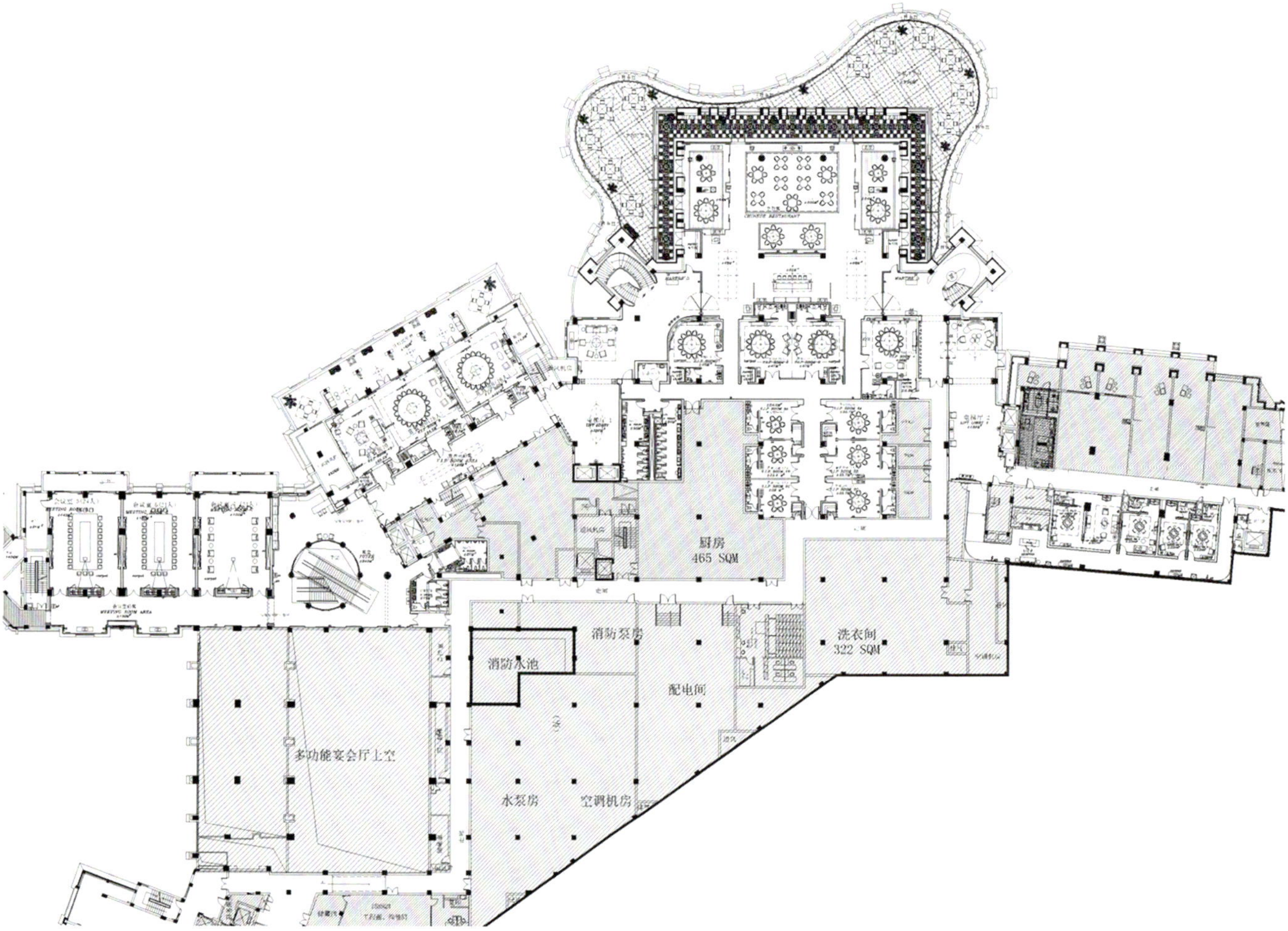
厨房
465 SQM
洗衣间
322 SQM
消防泵房
消防水池
配电间
多功能宴会厅上空
水泵房
空调机房

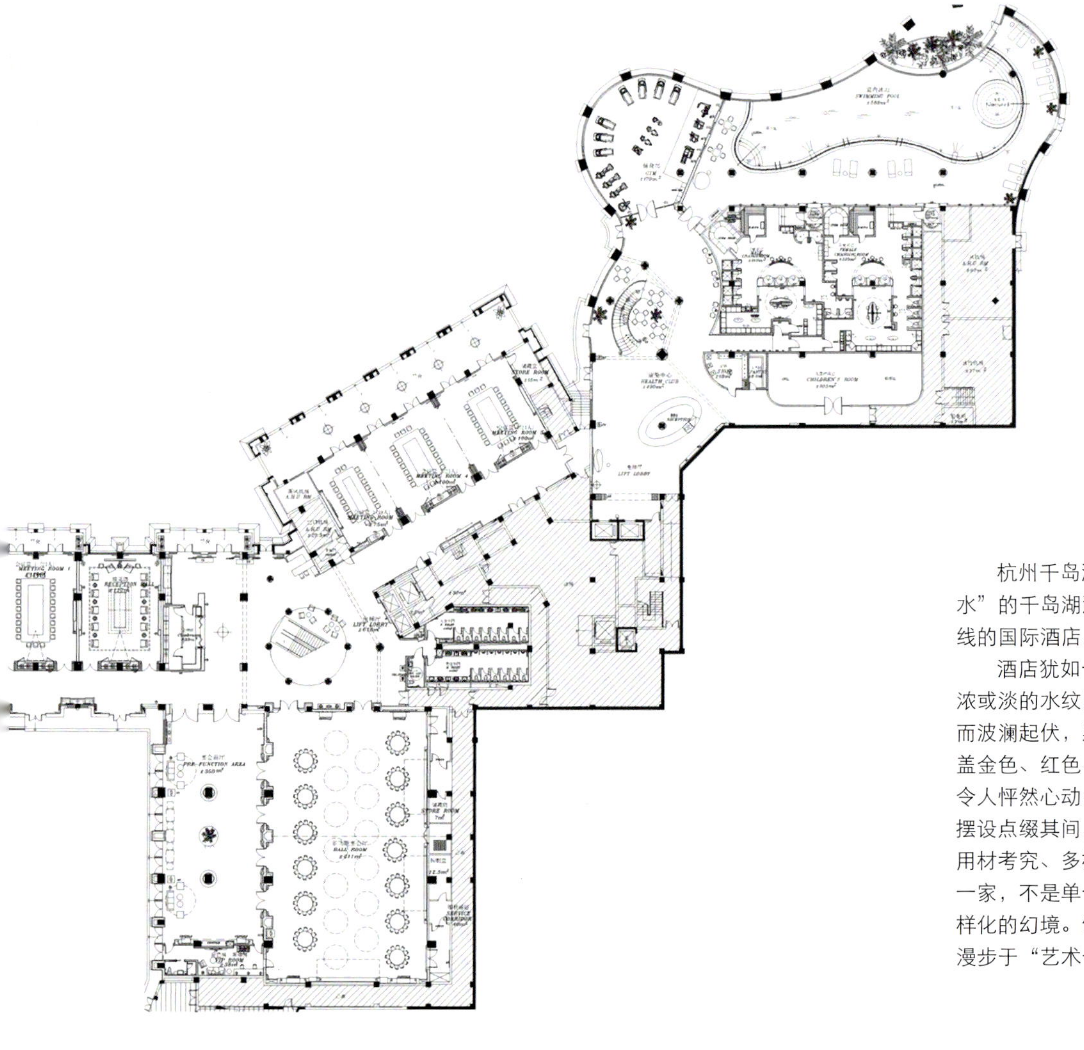

杭州千岛湖滨江希尔顿度假酒店位于“天下第一秀水”的千岛湖湖畔，依山面湖，是千岛湖拥有最长湖岸线的国际酒店。

酒店犹如一篇华美诗篇，以“水”起意，各区域或浓或淡的水纹图样，时而微波荡漾，时而涟漪浮动，时而波澜起伏，柔和、随性。设计如“水”包容万象，涵盖金色、红色、咖啡色等色调，个性、时尚元素凸显，令人怦然心动；新中式、简欧现代、东南亚色彩的艺术摆设点缀其间，时而高亢，时而低吟，韵律丰富多彩；用材考究、多样，带来多样化的格局冲击；风格也自成一家，不是单一的一种风格，多样化的格调赋予酒店多样化的幻境。信步其间，如诗般的文化底蕴飘然若现；漫步于“艺术长廊”，品味世间艺术万象，如梦如幻！

地点	面积	设计师	设计公司	主要材料
/ 福州	/ 9000m²	/ 黄炽烽、欧敏华	/ J2-DESIGN顾问设计有限公司	/ 紫檀木、柚木、灰木纹石等

Smart Hotel (Fuzhou)

时尚旅酒店(福州店)

The main theme of the hotel lobby is the biological shapes of wood and the walls are covered with combinations of dark and light colors of wood to match with the main color. The pillars are vertically wrapped with wood strips to resemble the tree trunk and become the focus of the space. Grey wood grain stone and wood floor are massively used to create a primitive and natural atmosphere of a forest. The furniture is decorated with jumpy colors and lively shapes to add highlights to the space so that the entire space is both alive and full of fun. The multi-functional hall on the second floor continues the style of the lobby on the first floor where vertical wood-strip shapes characterize the style of the space to reinforce the theme of the hotel and to create a harmonious and relaxing atmosphere at the same time, which makes the guests feel at home.

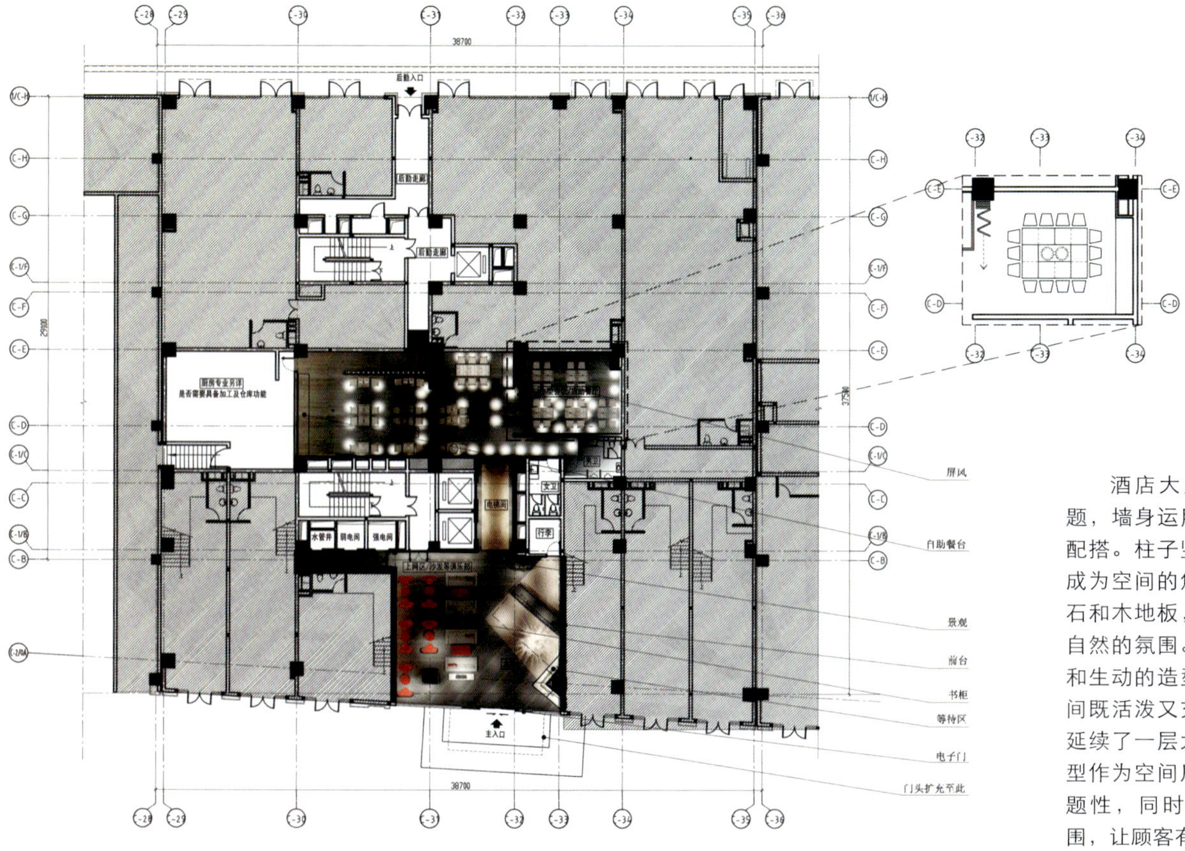

酒店大堂主要以木的生态造型为主题，墙身运用深浅木色的组合与主调进行配搭。柱子竖向以木条包裹，仿如树干，成为空间的焦点。地面大面积使用灰木纹石和木地板，整体营造出一种森林原始、自然的氛围。家具方面以点缀跳跃的颜色和生动的造型来增添空间亮点，使整个空间既活泼又充满乐趣。二层的多功能厅也延续了一层大堂的风格，以竖向木条的造型作为空间风格的延续，加强了酒店的主题性，同时也营造出一种和谐轻松的氛围，让顾客有种宾至如归的感觉。

地 点	设计师	设计公司
/ 广州	/ Manuel Derel, Eric Wong, Julian Cornu, Philippe Colin	/ 马迪思.明团设计机构

Qingcheng Hotel

倾城酒店

Qingcheng Hotel is situated beside the beautiful Yuexiu Mountain in Guangzhou, and near to Gaojianfu Memorial Hall. With east side next to Yuexiu Park, north side leaning against Museum of King of Southern Yue Tomb, and west side neighboring Liuhuahu Park, it offers us a superb location. With special intention, the hotel invites French pioneering designers for elaborate decoration, which provides it with favorable location, the right timing, and the most important thing is that it successfully creates the "harmony". Simple and elegant design and flowing area division produce something of exotic feelings in the hotel, which stands itself out in the whole city.

138 guest rooms with various styles are there in the hotel. Every room has a distinguished design and layout, but they all follow the principle of being "modern, fashionable, cozy and comfortable". On the premise that the general style is unified, different features are created in each individual room, which can supply different hotel accommodation experiences and it's the ideal choice for businessmen and travelers.

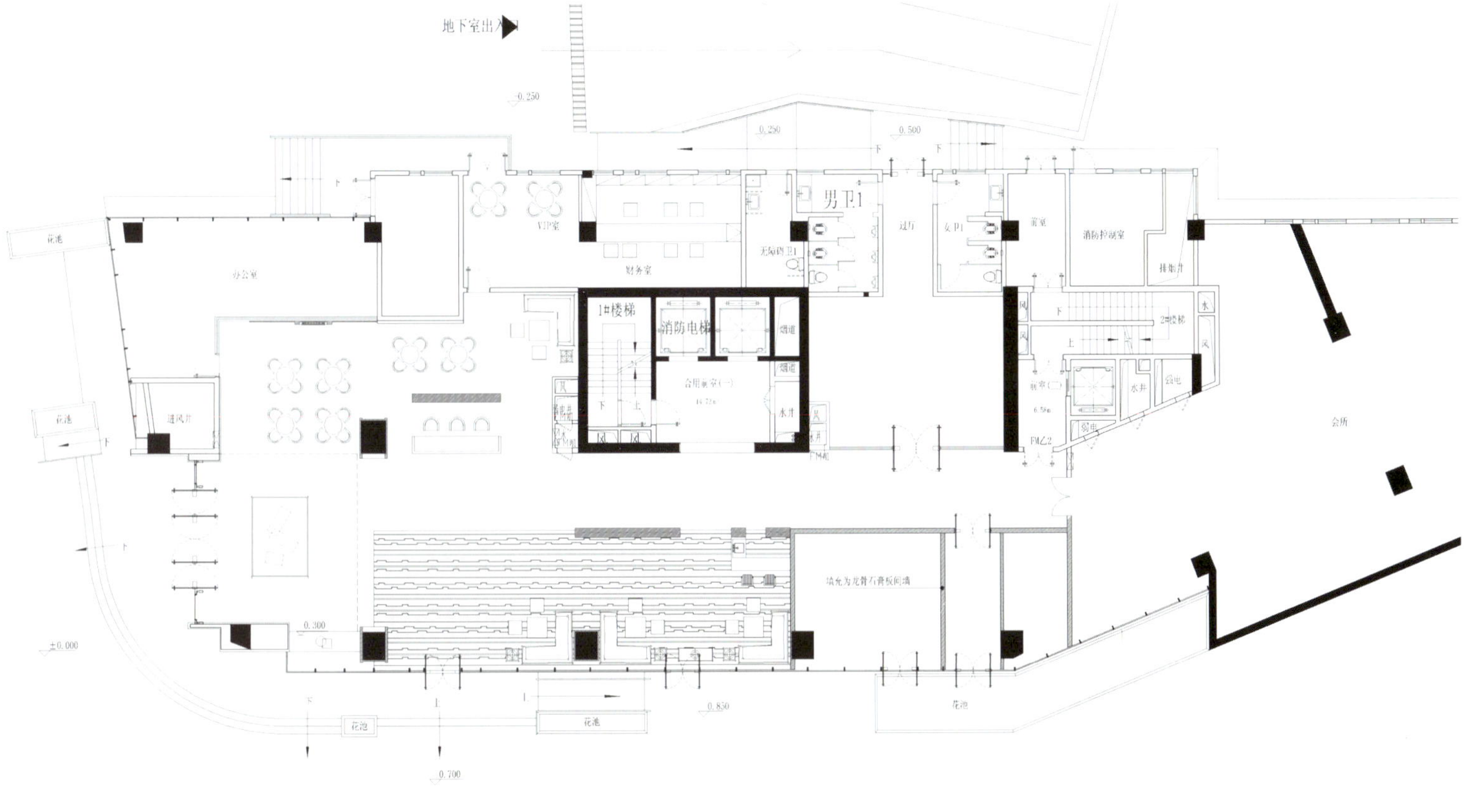
地下室出入口
男卫1
过厅
女卫1
消防控制室
VIP室
财务室
办公室
1#楼梯
消防电梯
合用前室(一)
2#楼梯
会所
花池
±0.000
0.250
0.500
0.850
0.700
0.300

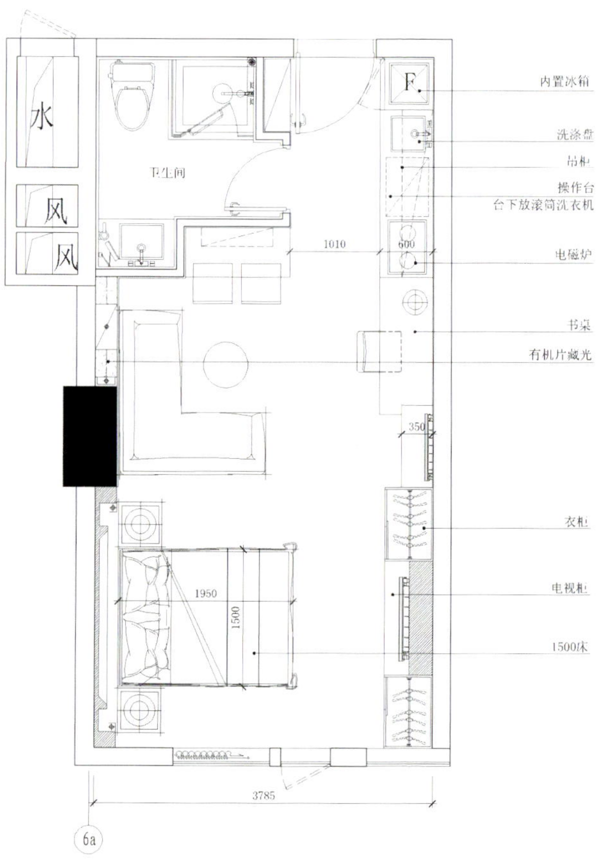

水
风
风
卫生间
F
内置冰箱
洗涤盘
吊柜
操作台
台下放滚筒洗衣机
1010
600
电磁炉
书桌
有机片藏光
350
衣柜
1950
1500
电视柜
1500床
3785
6a

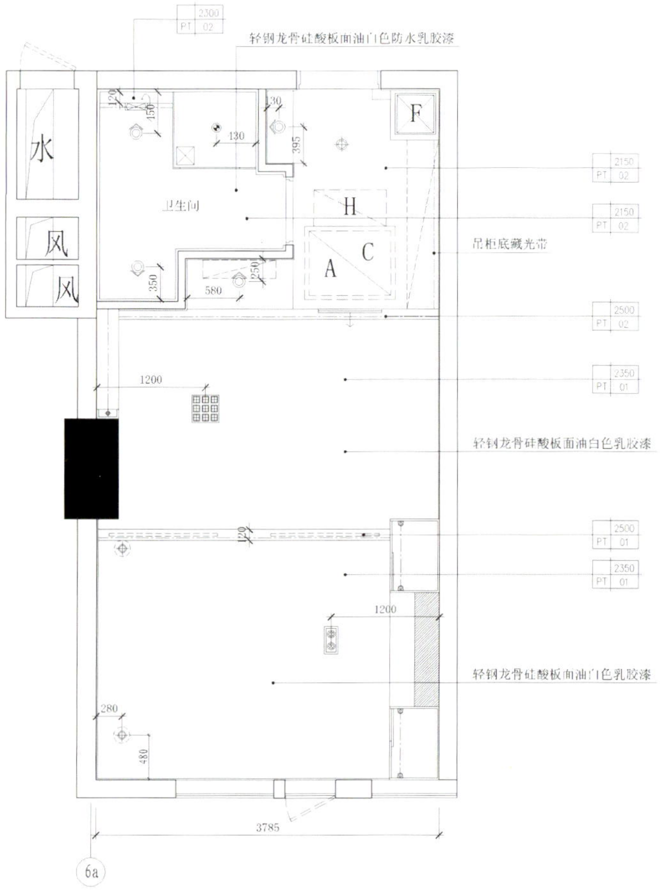

轻钢龙骨硅酸板面油白色防水乳胶漆
水
风
风
卫生间
F
H
A
C
吊柜底藏光带
130
395
150
350
580
250
1200
轻钢龙骨硅酸板面油白色乳胶漆
1200
轻钢龙骨硅酸板面油白色乳胶漆
280
480
3785
6a

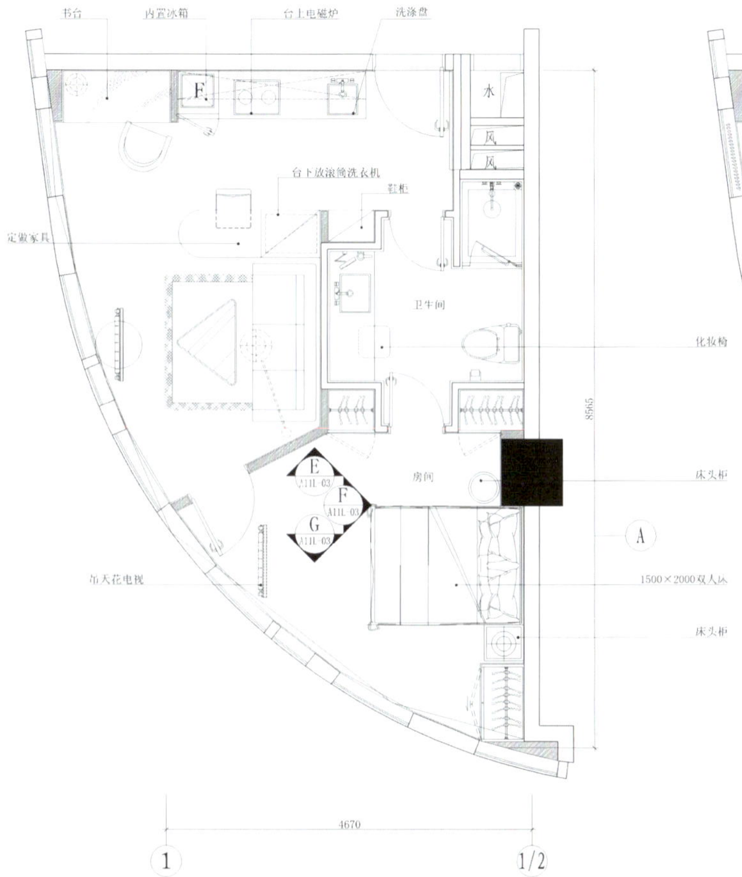
书台
内置冰箱
台上电磁炉
洗涤盘
台下放滚筒洗衣机
鞋柜
定做家具
卫生间
化妆椅
房间
床头柜
布天花电视
1500×2000双人床
床头柜
4670
8565
1
1/2

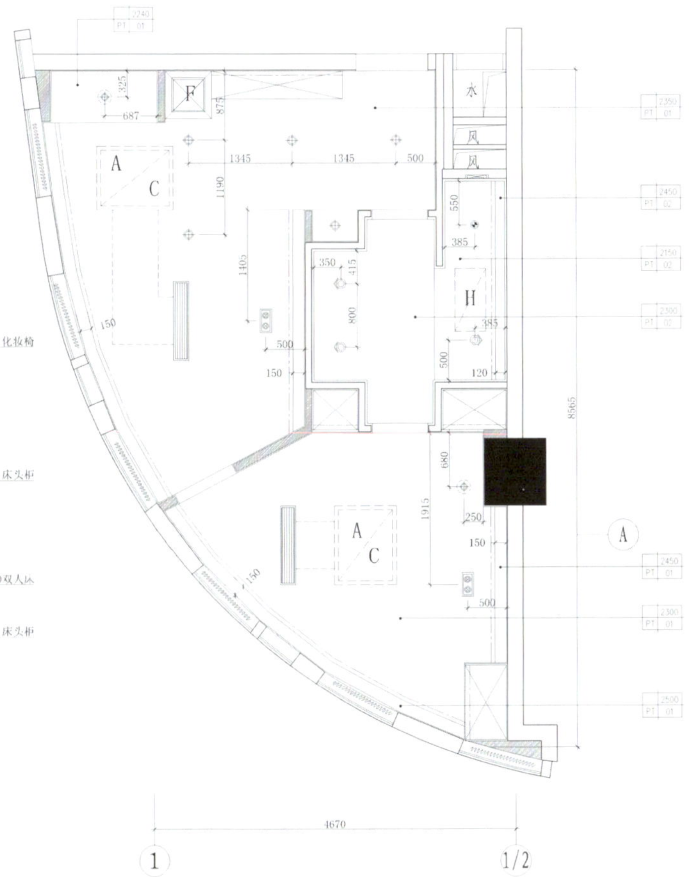
1345
1345
500
1190
1405
800
415
350
385
550
120
500
150
680
1915
250
150
500
687
4670
8565
1
1/2

150 Best Loft Ideas

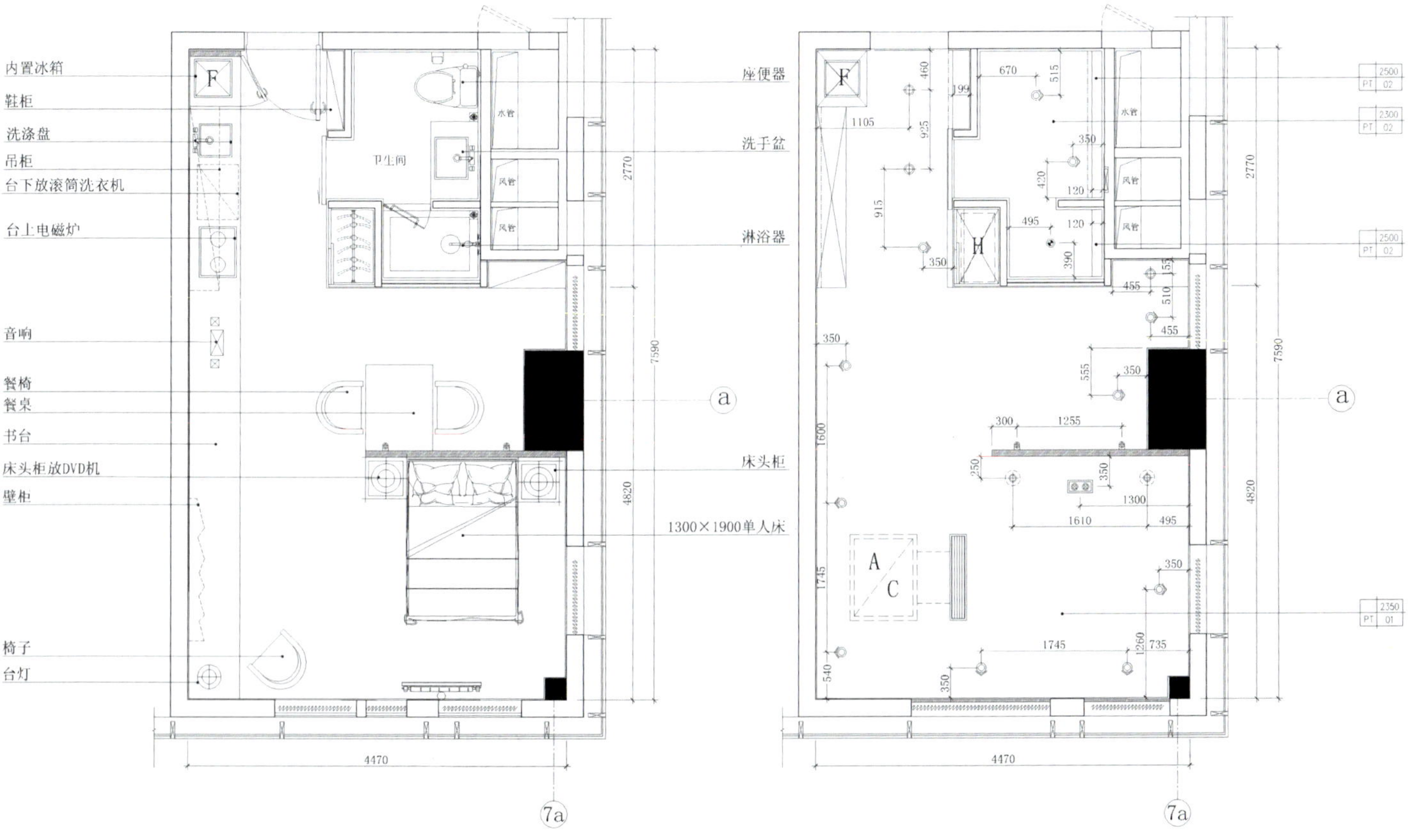
内置冰箱
鞋柜
洗涤盘
吊柜
台下放滚筒洗衣机
台上电磁炉
音响
餐椅
餐桌
书台
床头柜放DVD机
壁柜
椅子
台灯
卫生间
座便器
洗手盆
淋浴器
床头柜
1300×1900单人床
4470
7590
4820
2770

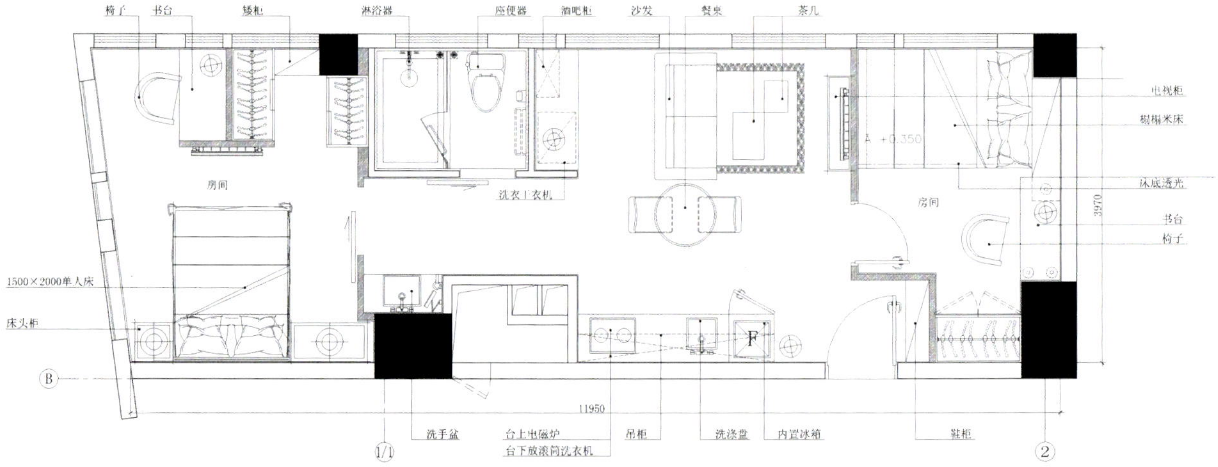
椅子
书台
矮柜
淋浴器
座便器
酒吧柜
沙发
餐桌
茶几
电视柜
榻榻米床
床底透光
书台
椅子
房间
房间
洗衣干衣机
1500×2000单人床
床头柜
11950
3970
洗手盆
台上电磁炉
台下放滚筒洗衣机
吊柜
洗涤盆
内置冰箱
鞋柜

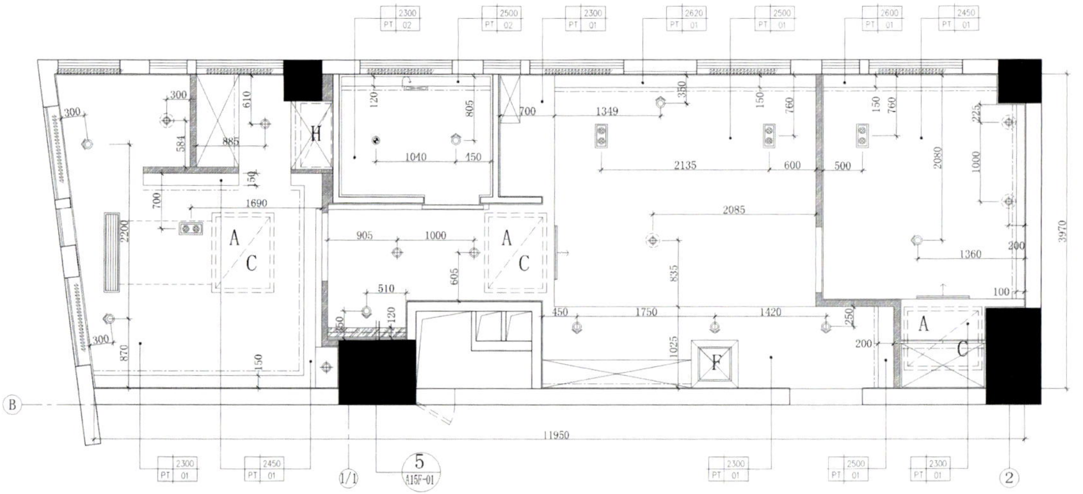
11950
3970
5
A15F-01

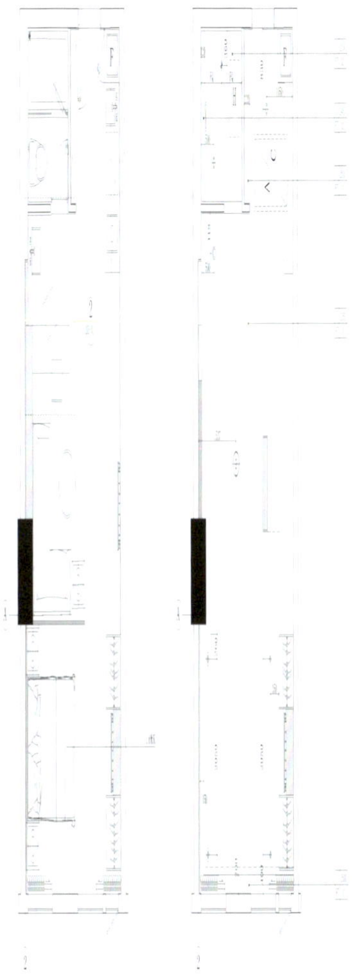

倾城酒店坐落于广州风景优美的越秀山旁，紧依高剑父纪念馆，东接越秀公园，北靠南越王墓博物馆，西邻流花湖公园，地理位置十分优越。酒店特请法国新锐设计师精心设计，让其不仅坐拥地利、天时，更要成功创造“人和”的条件，简约大方的设计及流畅的区域划分，让这个酒店生出几分异域风情，在都市中脱颖而出。

酒店共有138间风格各异的客房，每间客房设计布局各异，但都秉承着“现代、时尚，舒适、温馨”的原则，在大体风格统一的前提下创造各自不同的特色，让人体验不同的酒店住宿，是商旅人士的理想之选。

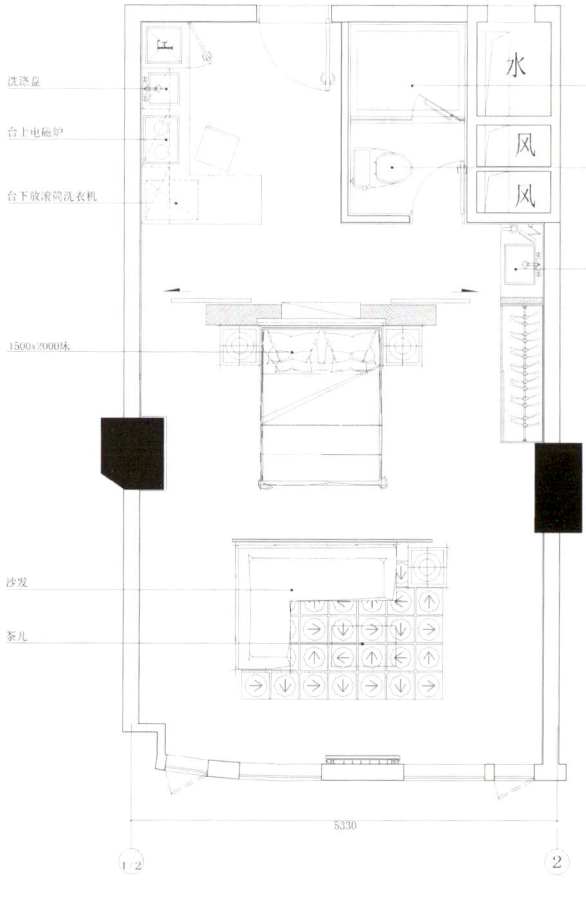
洗涤盆
台上电磁炉
台下放滚筒洗衣机
水
风
风
淋浴间
座便
洗手盘
1500×2000床
沙发
茶几
5330
1/2
2

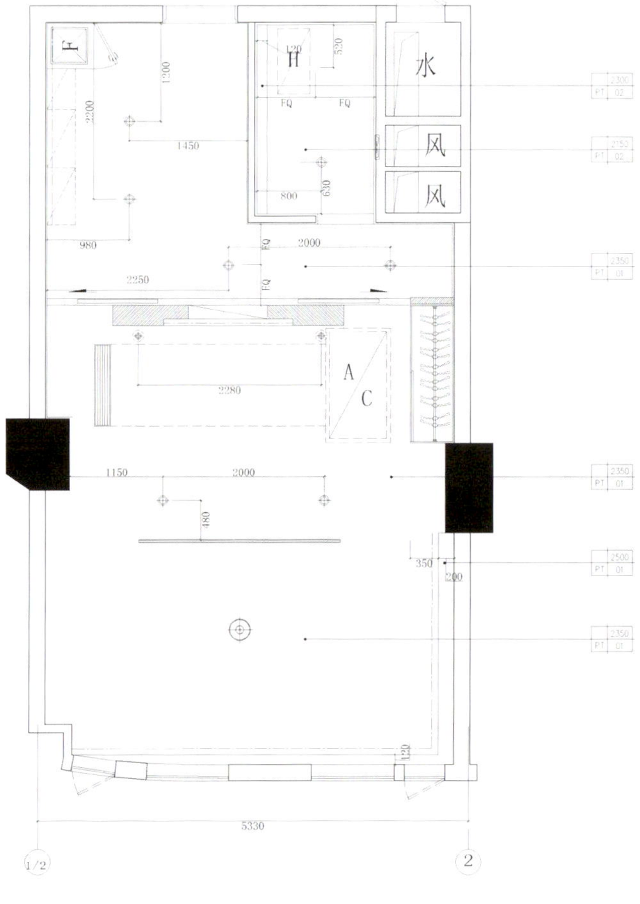
F
H
水
风
风
FQ
1200
2200
1450
800
630
980
2000
2250
2280
A
C
1150
2000
480
350
200
5330
1/2
2

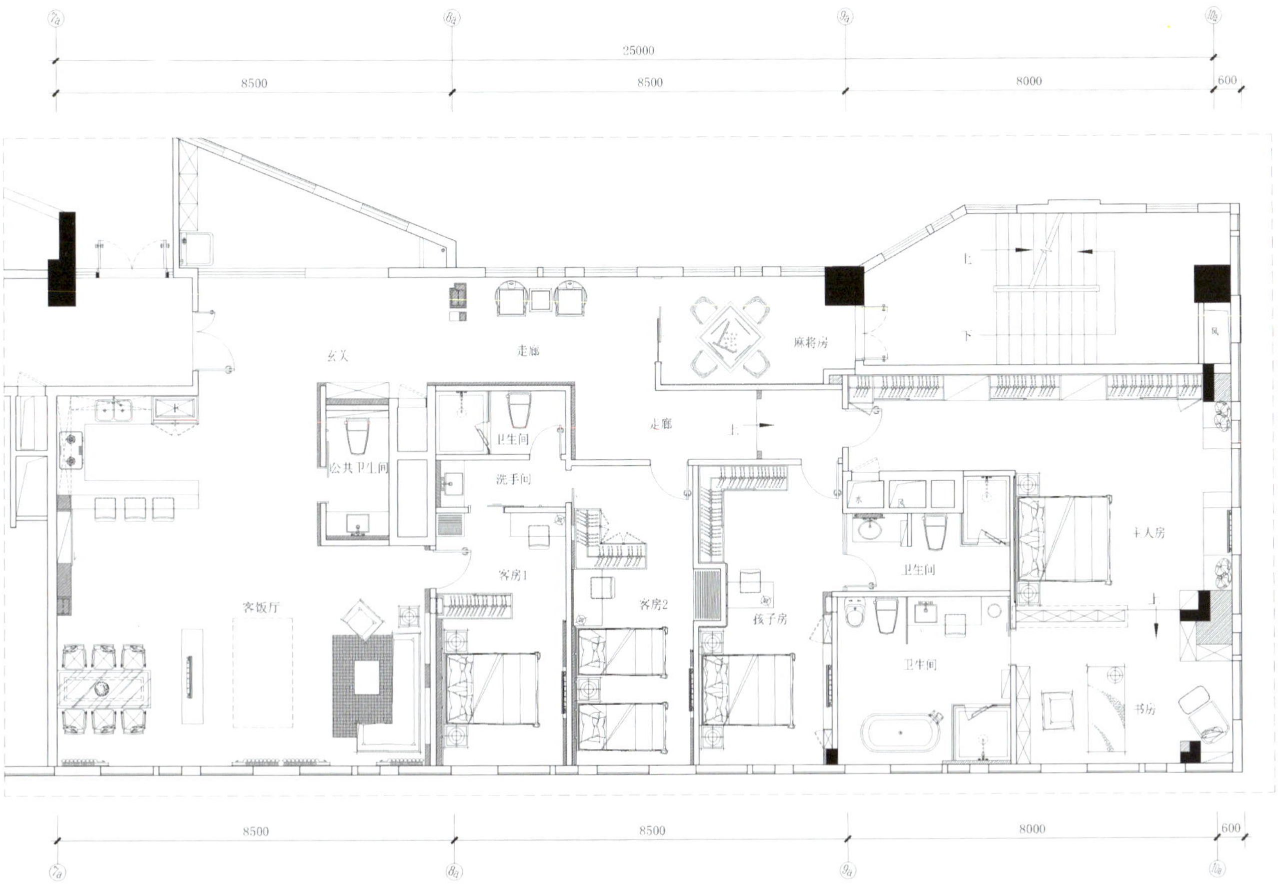

25000
8500
8500
8000
600
玄关
走廊
麻将房
走廊
上
卫生间
公共卫生间
洗手间
客房1
客房2
孩子房
卫生间
卫生间
十人房
书房
客饭厅
上
下
1300
2100
5400
11650
2850
300

本书参编人员

周锋、卢霭潮、刘宝达、欧阳亮、周强、陈哲、周美龄、雷小兰、胡青、吴俊、方丽、段君龙、周晓琪、庄丽娟、周琴、赵丹、赵标、闫兴宝、徐剑、王琪、黄芸、孙峰、黄宗坤、王雪松、贾春萍、李红靖、黄静、黄康裕、杜小慧、吴俭英